TRAVELLERS IN SPACE
AND TIME

TRAVELLERS IN SPACE AND TIME

Patrick Moore

Doubleday and Company, Inc.
Garden City, New York
1984

To Hilary Rubinstein with my grateful thanks

This book was designed and produced by
The Rainbird Publishing Group Ltd,
40 Park Street, London W1Y 4DE
for Doubleday & Company, Inc,
245 Park Avenue
New York, N Y 10167

First edition published in the
United States of America 1984

Editor: Caroline Zubaida
Designer: Lee Griffiths

ISBN 0–385–19051–4

Library of Congress Cataloging in Publication Data:
Moore, Patrick.
    Travellers in space and time.
    Includes index.
    1. Astronomy – Popular works. 2. Space flight –
Popular works. I. Title.
QB44.2.M692 1984   520
ISBN: 0-385-19051-4
Library of Congress Catalog Card Number: 83-7401

The text was set, printed and bound and the colour
originated by Jarrold & Sons Ltd, Norwich, England.

Half title: The Jovian system. Jupiter is attended by its four main
satellites: Io, Europa, Ganymede and Callisto. The pictures were
taken by Voyager 1.

Frontispiece: The Great Nebula in Orion; we will make a closer
survey later, and meanwhile we will do no more than note the
form and the bright colours of this immense cloud of dust and
gas which is nothing more nor less than a stellar nursery.

Contents

Overleaf: Journey through space: starting from the Earth-Moon system, our journey takes us first to the inner planets prior to leaving the Solar System altogether. We travel to the nearby stars and head for the heart of the Galaxy. We then virtually retrace our steps and leave the Galaxy, visiting the major members of our local group of galaxies. Still deeper in space, we pass the giant clusters of galaxies which dwarf our own local groups. Finally we travel to the very limits of the observable universe. (The distances are given in light hours and light years.)

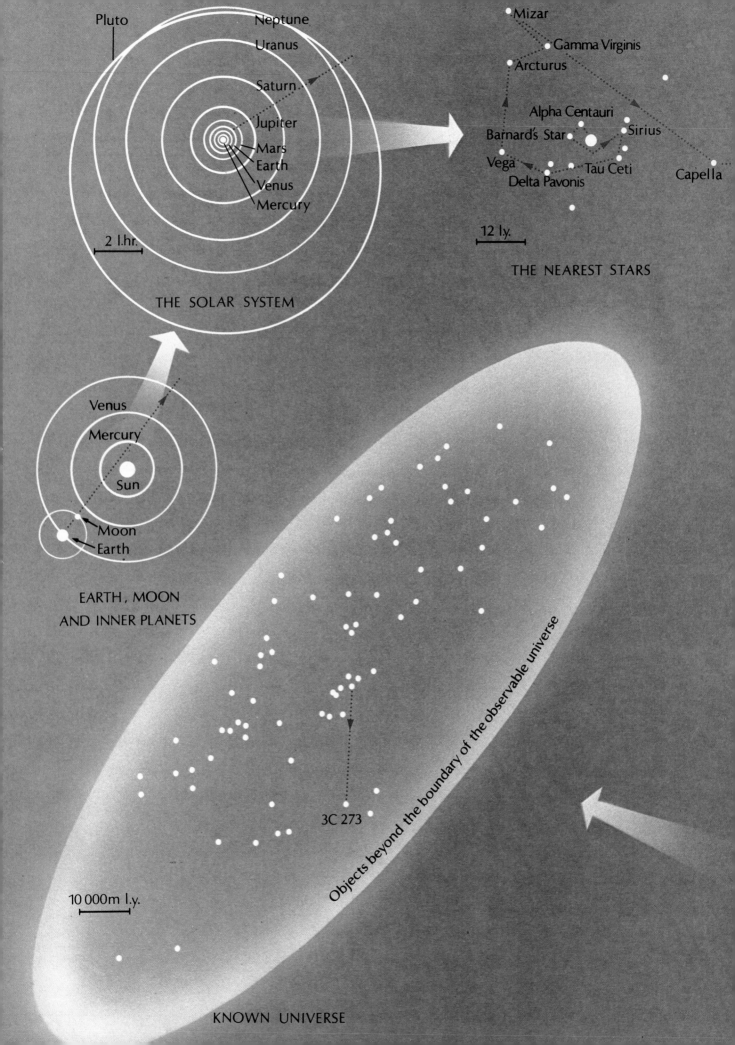

Pluto

Neptune

Uranus

Saturn

Jupiter

Mars
Earth
Venus
Mercury

2 l.hr.

THE SOLAR SYSTEM

Mizar

Gamma Virginis

Arcturus

Alpha Centauri

Barnard's Star

Sirius

Vega

Tau Ceti

Delta Pavonis

Capella

12 l.y.

THE NEAREST STARS

Venus

Mercury

Sun

Moon
Earth

EARTH, MOON
AND INNER PLANETS

Objects beyond the boundary of the observable universe

3C 273

10 000m l.y.

KNOWN UNIVERSE

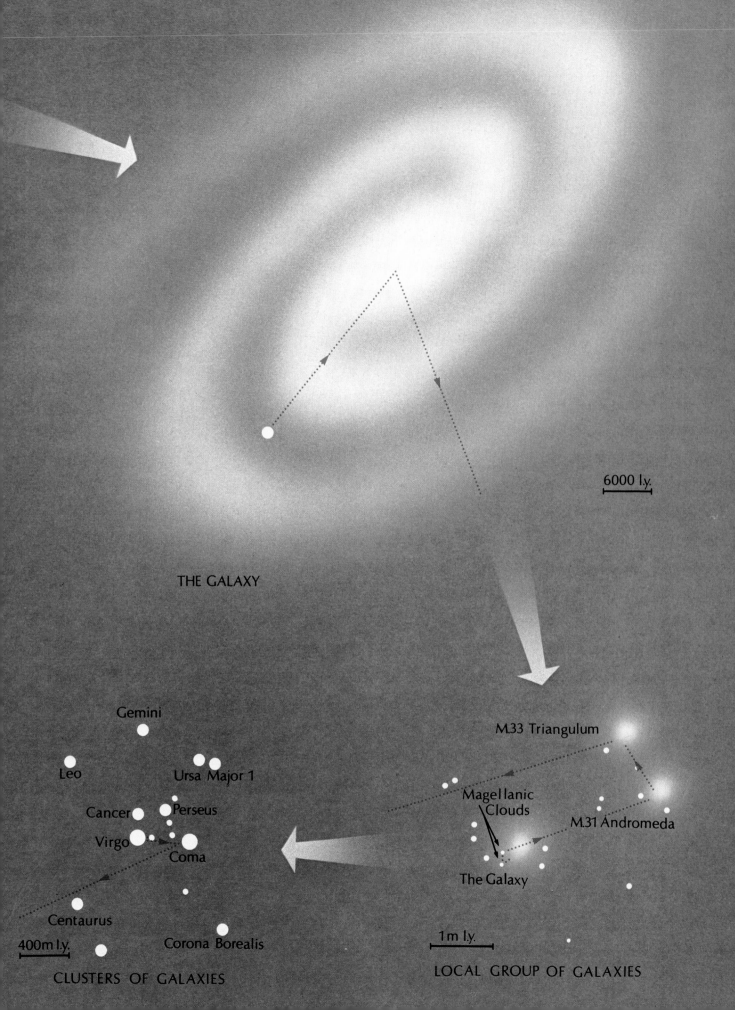

6000 l.y.

THE GALAXY

Gemini

Leo

Ursa Major 1

M.33 Triangulum

Cancer

Perseus

Magellanic
Clouds

Virgo

M.31 Andromeda

Coma

The Galaxy

Centaurus

1 m l.y.

400m l.y.

Corona Borealis

CLUSTERS OF GALAXIES

LOCAL GROUP OF GALAXIES

1 At the Speed of Light

Far away in space, so remote that it looks like nothing more than a dim glow, lies the system we call the Andromeda Spiral. It has been known for over 1,000 years; the old Arab star-gazers saw it, but not until modern times did we find out its true nature. It is made up of stars – more than 100,000 million of them – and it is so far away that even its light, travelling at 186,000 miles per second, takes over 2,000,000 years to reach us.

Light moves faster than anything else in the universe, but it does not travel instantaneously. Switch on a torch in a darkened room, and there will be a definite interval before the beam reaches the opposite wall, though the delay is so slight that it could certainly not be measured. Light takes one and a quarter seconds to reach us from the Moon, which is much closer to us than any other natural body in the sky. This means that we see the Moon not as it is now, but as it used to be one and a quarter seconds ago. This is trifling; but with the Andromeda Spiral, the situation is different, and our present-day view of it is very out of date, because the light we are just receiving from it started on its journey towards us over 2,000,000 years ago, well before the beginning of the last Ice Age.

Each star is a sun. Probably not every sun has a system of planets, but there is every reason to believe that 'other Earths' must be common in the universe, in which case there

The start of our journey: the Andromeda galaxy. An inhabitant of a planet inside the great spiral system would see our Galaxy from a distance of over 2,000,000 light-years, and would have little hope of identifying a feeble star like our Sun.

will be plenty of them in the Andromeda Spiral. Suppose that we give our imagination full rein, and transfer ourselves to a planet inside the Spiral; what can we expect to see if we look in the direction of the system which contains our own Sun?

If our Andromedan has eyes no better than ours, the answer will be 'nothing'. The Andromeda Spiral is only just visible from Earth with the naked eye, and it is larger than our Galaxy, so we must equip our Andromedan with binoculars. Then he will be able to make out a faint blur of light, of no definite shape and with no definite starlike points inside it. It will seem unimpressive, and not nearly so bright as two other systems which are much closer to him.

Next, assume that the Andromedan astronomer is able to use a telescope as powerful as anything we have yet built on Earth. Now he will see real detail; the shape of the system will be revealed as a rather loose spiral, like a Catherine-wheel, with dark 'lanes' here and there which he will know to be made up of dusty material, not lit up by any suitable star. He will also see a great many individual stars, and if he is equipped with instruments similar to our spectroscopes, which split up light and indicate which materials exist in the light sources, he will realize that he is looking at a system very similar to his own apart from the fact that it is decidedly smaller.

If he keeps on watching for a sufficiently long period of time, he and his fellow Andromedans will see changes. Some of the stars in our Galaxy brighten and fade

Planet Three – the Earth. It is not too easy to make out the shapes of the continents and oceans; there are too many clouds, but at least the form of Africa can be seen. It is easy to understand why the Earth is often called the Blue Planet.

supernovæ. But our Andromedan may have to wait for a long time, because supernovæ are rare. The last supernova to be seen in our own Galaxy was that of the year 1604, but the light from it will take more than 2,000,000 years to reach Andromeda.

This may be about as much as our alien astronomer can find out. However, since we are travelling in imagination, we can fasten ourselves on to a ray of light and start the long journey between Andromeda and the Earth. About 1,000,000 years after departure, the distance will have been halved, and the visitor will see the spiral arms of our Galaxy very clearly. More and more individual stars will come into view; not only very powerful giants, some red and others bluish or white, but also more ordinary, run-of-the-mill stars, many of which are yellowish in colour. With sufficient foreknowledge, our Andromedan will be able to concentrate his attention upon one particular yellow star, well away from the centre of the system and just outside the edge of one of the spiral arms. This is the star we call the Sun, though from 1,000,000 light-years there will be nothing to single it out from its neighbours.

However, other features are becoming evident. The Galaxy is not alone. Adjoining it are two other systems, one bigger than the other, although both are much smaller than the Galaxy itself. We call them the Clouds of Magellan. Both lie within 200,000 light-years of the Earth – one light-year being the distance travelled by a ray of light in one year, equivalent to rather less than 6,000,000 million miles. The Clouds include stars of all kinds, as well as dusty material; they are irregular in outline and are contained inside a vast cloud of highly rarefied gas.

As our Andromedan draws nearer and nearer, the picture will change. At last he will reach the outer edge of the Galaxy, about 50,000 light-years from the centre. He will not be able to see through to the centre itself, since there is too much obscuring

regularly; these are the variable stars. Some of them give away their distances by the way they behave, so that it is possible to work out the distance between Andromeda and the system in which the variable stars lie. Occasionally, there will be a flare-up, when a formerly obscure star explodes in a blaze of light that lasts for several weeks or months. Outbursts of this kind are termed material in the way, but he will be able to guess where it is. He will also know that the Sun, to which he is travelling, is about 33,000 light-years from the galactic centre, and lies almost in the main plane of the system. Another 17,000 years of travel on his light-beam, and he will be closing in.

He will find that the Sun is very isolated. No other star exists within four light-years of it, corresponding to a distance of around 25,000,000 million miles, but at this range there is a trio of stars between 4.2 and 4.3 light-years away, making up what we call the Alpha Centauri group. Two of its members are bright, one of them slightly more luminous than the Sun and the other slightly fainter; the third member, Proxima, is a dim red star, a glow-worm on the cosmical scale. But by now our Andromedan will be coming within range of the Sun, and the tiny dot of light will have become a brilliant yellow disk.

He will also have found that the Sun is not a lone traveller. Circling it are several much smaller worlds, which have no light of their own and which shine simply by reflecting the rays of the Sun in the manner of mirrors. The Andromedan will decide that this is a true planetary system, and will also see that it is divided into two parts. First to come into view as he draws nearer will be the relatively large planets which we call Neptune, Uranus, Saturn and Jupiter; if he has been able to bring his spectroscope he will find that their surfaces are gaseous rather than solid, and he will see that all of them are attended by junior bodies or satellites. One of the planets, Jupiter, is much more massive than all the rest, and for a moment our visitor may wonder whether it can be in the nature of a faint star. Closer inspection will show that this is not so. Jupiter is indeed large – over 88,000 miles in diameter, and big enough to swallow up over 1,000 bodies the volume of the Earth – but it is not self-luminous. If the Sun were to

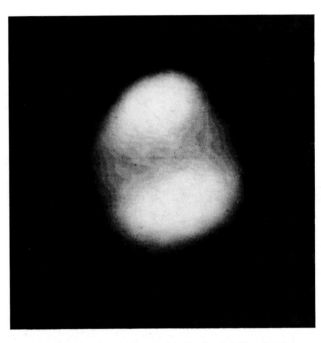

Neptune. The outermost giant; its bluish clouds hide what lies below, and as we pass by, it is possible to see no more than rather vague, cloudy patches in Neptune's upper atmosphere.

disappear suddenly, then Jupiter and the other planets would vanish too.

Now the Andromedan can cut down his speed. If he did not, he would pass through the inner part of the Solar System in a few hours. He decelerates until he is moving at only a few thousands of miles per hour, and can take stock of the planets closer in to the Sun. There are four in number; reckoning outwards, they are Mercury, Venus, the Earth and Mars. This time we have bodies with solid surfaces. All except Mercury have appreciable atmospheres, and in the case of Venus the atmosphere is so dense and so cloud-laden that the true surface is completely hidden. Our Andromedan may wonder what conditions are like below the clouds; can there be oceans, lands and living creatures? But it is Planet Three which catches his attention. It is a blue world, and it is easy to make out the shapes of continents and seas. Moreover there is a companion world, the Moon, which has a barren, crater-scarred surface and gives every impression

Venus from Pioneer; it looks beautiful, but there are no details visible in its upper atmosphere, and nothing to warn us of the inferno-like conditions existing below the cloudy mantle.

of being dead. It could even be described as a cosmic museum.

Unless our Andromedan approaches the Earth to within a few thousands of miles, he will be unable to make out any signs of artificial constructions or buildings, but at least he is bound to appreciate that here, perhaps for the first time during his journey, he has come across a world which is ideally suited to life. He may linger; but let us assume that he is more interested in the Sun, the central body and ruler of the Solar System.

Here we have a globe which is truly awe-inspiring. The surface is hot; the gases are at a temperature of nearly 6,000 degrees centigrade, and no rocket built of ordinary materials could approach it closely. Luckily our Andromedan is not so limited, and he can swoop down through the extensive, incredibly thin solar atmosphere, which we call the corona, to hover above the brilliant disk itself.

The bright surface, or photosphere, is not calm and placid, but is in a state of constant

Saturn. Seen from close range the Ringed Planet is truly magnificent. The yellow globe, with its gaseous surface, casts a shadow across the rings; we can see that the rings themselves are made up of many minor ringlets and gaps.

turmoil. Jets of gas rise from below, spread out and vanish, to be replaced by new ones; the whole surface is pulsating, and here and there may be seen darkish patches or sunspots, which are some 2,000 degrees cooler than the photosphere and which therefore look black (though if they could be seen shining on their own, their surface brightness would be greater than that of an arc lamp). Moreover, the Sun is sending out radiations of all kinds. As well as visible light, there are short-wave radiations, ranging from ultraviolet through to X-rays and gamma-rays; there are also long-wavelength radiations, which are known as

radio waves, though there is no suggestion that they are artificial.

At this point, our Andromedan may pause to consider. When he approached the Solar System to within a distance of about 60 light-years, he had started to pick up different kinds of radio waves, not in the least like those from the Sun or various other natural sources which he had bypassed during his journey. Beyond 60 light-years these signals had been inaudible. What can be the reason? He can work it out. These signals were indeed artificial, coming not from the Sun but from one of its planets, presumably the friendly looking Planet Three. If the broadcasts began only 60 years ago, they would have penetrated no more than 60 light-years into space. Beyond that, the Earth is radio-quiet; closer in, it becomes

Impression of Uranus from close range. The Green Planet reveals few of its secrets even when we approach it; but we can see some cloudy structures and also two of its satellites, Miranda (left) and Ariel (right). The faint ring system of Uranus is visible.

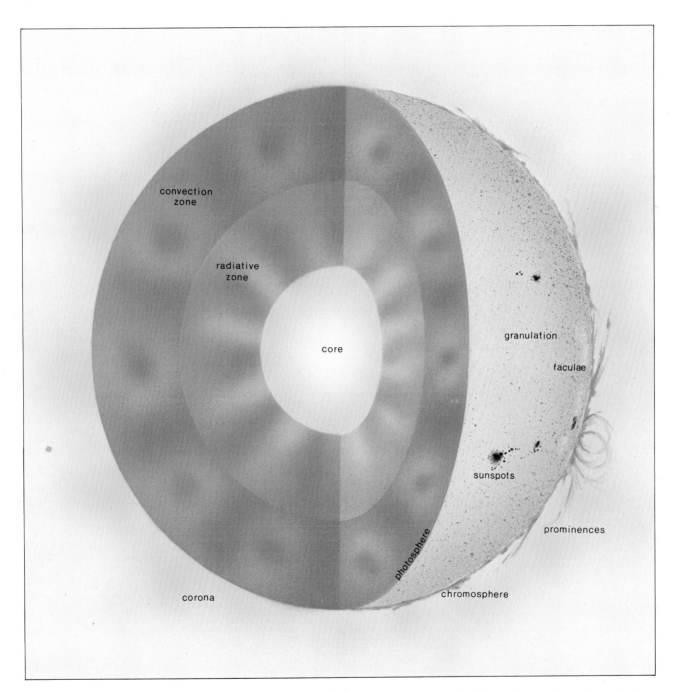

Travel into the Sun; we pass the bright outer atmosphere – the prominences (once called Red Flames) and the 'colour sphere' or chromosphere, before plunging down into the solar core, where the Sun's power-house is situated.

radio-noisy. The inescapable conclusion is that Planet Three supports an intelligent civilization which has started to transmit.

By now the Sun is close ahead, and the intense heat would destroy any material object as effectively as a blast-furnace will destroy an insect unwise enough to fly into it. Our Andromedan need have no such qualms; immune to heat and pressure, he can plunge through the photosphere and deep into the Sun's globe. Steadily the temperature rises, and the whole environment changes. The photosphere is made up of fundamental substances or elements which the Andromedan will know as well as

we do; hydrogen, helium and the rest, with a tremendous excess of hydrogen over all the other elements combined. But when the temperature soars to millions of degrees, the atoms are broken up, and we are left with their debris.

For most purposes it is good enough to say that atoms, which make up all the material in the universe, may be regarded

Jupiter from Voyager. As we approach, we can see that it is brilliantly coloured; the upper clouds are in constant turmoil, and the Great Red Spot, a whirling storm (towards the bottom of the picture), dominates the scene.

as miniature Solar Systems. In the centre there is a nucleus, made up of protons and neutrons; a proton carries a unit charge of positive electricity, while a neutron has no charge at all. (Hydrogen, the simplest of all atoms, has a nucleus consisting of a single proton.) Around the nucleus move planetary electrons, each of which carries a unit charge of negative electricity. In a complete atom, the numbers of planetary electrons exactly balance the combined charge of the protons in the nucleus. Thus the nucleus of an atom of oxygen contains 16 protons; there are 16 planetary electrons, and the overall charge is nil, since $+16 - 16 = 0$. But deep inside the Sun, all the electrons have been wrenched away from their nuclei, and there is nothing more than a disorganized hurly-burly of atomic fragments.

Yet the mêlée is not entirely chaotic. If four hydrogen nuclei (that is to say, four protons) run together, they combine to produce a nucleus of the second lightest element, helium. Each time this happens a little energy is released, and a little mass (or 'weight', if you like) is lost. It is this energy which keeps the Sun shining, so that hydrogen acts as the all-important solar fuel.

By now our traveller has reached the core of the Sun, some 430,000 miles below the surface. It is just as well that he is impervious to heat; if he carries a thermometer he will find than it registers at least 14,000,000 degrees centigrade, probably rather more. The material is extremely dense, but it is still a gas, and behaves like one. We are in the Sun's power house.

Our Andromedan will pause once more, and think hard. So the Sun, like all normal stars, is not burning in the usual sense of the term; its energy is due to nuclear reactions, the changing of hydrogen into helium. The process is not entirely straightforward, but this need not bother him for the moment, because at least he can understand the general picture.

He can decide, too, why the Sun is so different from a planet, even a large planet such as Jupiter. Inside Jupiter the temperature may well be as high as 30,000 degrees centigrade, but this is not nearly enough to spark off the hydrogen-into-helium reaction. We need at least 10,000,000 degrees, and the Sun's core is much hotter than that.

The energy produced near the Sun's core is of very short wavelengths, and takes millions of years to percolate through to the outer surface before escaping into space, by which time it has been changed into visible light. But there are other things, too: neutrinos, which have no electrical charge and no mass (or, at least, very little; some doubts have been expressed recently). Neutrinos can streak out, passing through the Sun's globe as easily as a beam of light can pass through a glass window, and therefore they provide the only means of obtaining up-to-date news from the Sun's power house, short of making an actual journey inside the globe.

There is another factor which our Andromedan must bear in mind. The Sun is shining because it is using hydrogen as a fuel, but the supply of fuel cannot last for ever. Sooner or later it will be exhausted, and the Sun will be forced to change its structure. The present loss of mass amounts to 4,000,000 tons every second, but cosmically speaking this is almost negligible, and a few quick calculations will show that there is enough hydrogen fuel to keep matters much as they are for at least 5,000 million years in the future. It also follows that since the Sun is now middle-aged, it has existed, more or less in its present form, for about 5,000 million years. Our Andromedan thinks hard. Suppose he could look at the Solar System from a distance of, say, 8,000 million light-years. Would he see the Sun? No, because his view would date back before the time when the Sun began to shine.

The journey has reached its climax, and it is time to return home. Travelling at the speed of light, our visitor leaves the solar core; within three seconds he is back outside the photosphere, and begins to draw back into the depths of space, still maintaining his speed. No time now to pause and take a closer look at the planets; a few hours sees him arrive at the edge of the main Solar System, marked by the path or orbit of the giant planet Neptune, and then he is back in interstellar space. Yet he has not quite left the Solar System behind. Beyond Neptune there are flimsy, ghostlike bodies called comets, some of which invade the inner part of the Sun's family while others keep prudently out of range. A year or so after starting his return, our Andromedan passes through a huge swarm of comets called the Oort Cloud, named in honour of the Dutch astronomer Jan Oort, who first suggested its existence. There are millions of comets in the cloud, but they are very dim since they pick up so little sunlight, and they are relatively slow-moving. Next even the Oort Cloud is left behind; four years brings the traveller back out to Alpha Centauri, and the Sun has once more shrunk to a pinpoint of light, while all its planets, even Jupiter, have been lost to view.

Can our Andromedan know anything about the past history of the Earth? Assume that he can, and assume, too, that he is carrying a telescope powerful enough to show the Earth's surface in great detail. By the time he has receded to 900 light-years he may have come within range of the brilliant white star Rigel in the constellation of Orion, which is at least 60,000 times more powerful than the Sun. He trains his telescope on Earth: what does he see? No cars, no blocks of flats, no motorways or airports. To him, the Earth is seen as it used to be 900 years ago. If he can concentrate upon England, he will be able to see the armies of William the Conqueror as they move around, carrying out their task of subduing the Saxons. As he speeds on and on, he reaches another bright star, Deneb, at a distance of 1,800 light-years, and his telescope shows Britain at the time of the Roman Occupation. But without such a miraculous telescope he will learn little more; and as he recedes further and further he will lose first the individual stars of the Galaxy, and then its spiral form. He will arrive back at his home planet in Andromeda very much the wiser for his expedition, but he will also realize that he has covered only a tiny fraction of the whole of the universe.

Of course, travelling at the velocity of light is something which is quite beyond us at the moment. Indeed, so far as material objects are concerned it may never be achieved. Einstein's theory of relativity, published early in our own century, lays down some very curious effects which start to make themselves felt when an object is moving at very high speed. In particular, the traveller's time scale is slowed down. If he could move at, say, 99 per cent of the velocity of light, he could travel to Alpha Centauri and back in a period which would seem to him to be about ten years; but on his return to Earth he would find that he had been away for much longer than ten years because his time scale, relative to that on Earth, has been changed. Worse, one's mass increases at these tremendous speeds. If Einstein is right, then a body travelling at the full 186,000 miles per second would find that its mass had become infinite and its time scale had come to a complete stop, which is another way of saying that it can't be done. This assumes, of course, that there is no major flaw in relativity theory. It is always dangerous to be dogmatic, but it is fair to say that every test so far applied has ended in

Mars. We can already begin to see the huge volcanoes, Mount Olympus the largest volcano anywhere in the Sun's family. We can also make out the white patch covering the Martian pole.

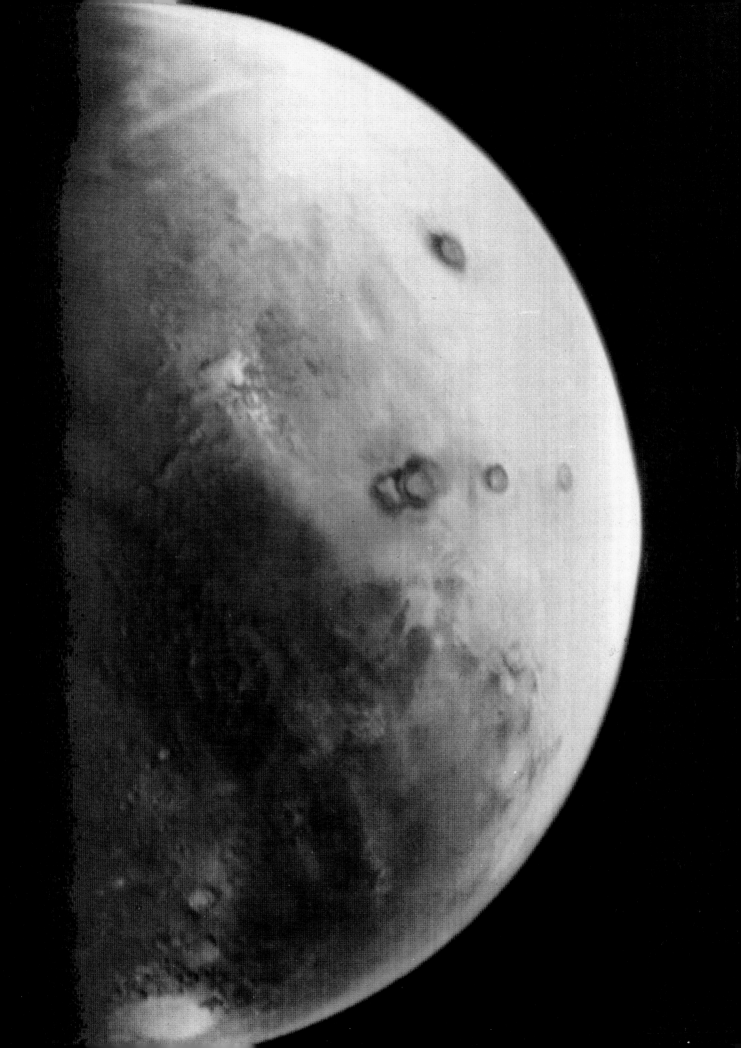

Einstein's favour and the discomfiture of his critics. The 'time-dilation effect' can even be checked experimentally. The Earth is under constant bombardment by cosmic rays, which are not really rays at all, but high-velocity atomic nuclei coming from all directions in space. As the nuclei hit the Earth's upper air, they smash into the air particles and are broken up; the air particles are broken, too, and particles called mu-mesons are created. They last for so short a time before decaying that they ought not to survive for long enough to reach the Earth's surface; yet they do, because their time scale, relative to ours, has been slowed down. Recently there has been an even more dramatic check. A highly accurate clock was flown round the Earth in a jet aircraft, travelling at its maximum possible speed. This is very sluggish when compared with the velocity of light, but it was enough to show that when the experiment was over the travelling clock had 'lost' a tiny amount of time compared with a similar clock which had been kept in a laboratory on the ground.

A rocket of the type which we can build today can take a man to the Moon in a couple of days, and can reach the outermost giant planet, Neptune, in a dozen years. But to reach even Alpha Centauri would take many centuries, and there is no way either of sending out a manned expedition or of keeping track of an automatic probe out to such a distance. Indeed, at this very moment there are four spacecraft, two Pioneers and two Voyagers, which are on their way out of the Solar System. They will leave the main part of it during the 1990s, but before the year 2000 we will have lost track of them, and we will never know their ultimate fate. They will travel on, unseen and unheard, perhaps for thousands of millions of years, until either they are destroyed by collision with some solid object or (much less probably) collected by an alien civilization

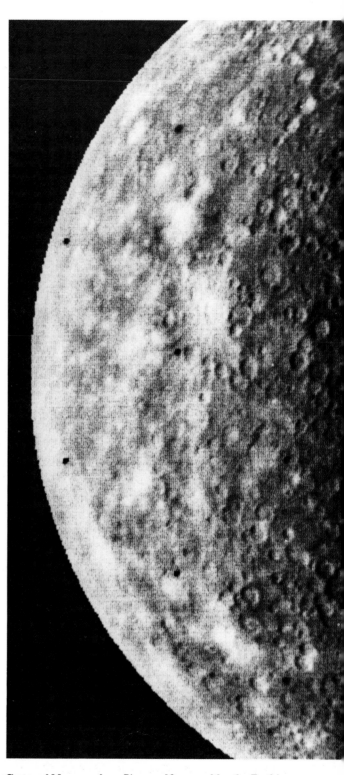

Cratered Mercury, from Pioneer 10, resembles the Earth's Moon, though we look in vain for the wide, relatively smooth plains which make up the lunar 'seas'.

and removed to a museum. Each carries a plaque which will, it is hoped, give a clue to its planet of origin, but nobody pretends that this is anything but the longest of long shots.

Therefore, interstellar travel by present-day methods is impossible. This is not to say that it will always be impossible; after all, William the Conqueror would have had little patience with anyone who told him that within a thousand years after his death it would be possible to sit at home, turn a switch, and watch a man walking about on the surface of the Moon. But we need a fundamental breakthrough, and we cannot tell when this will be made – if it ever is.

What, then, are the chances that some other civilization in the Galaxy has already made this breakthrough and has sent messengers to the Earth? Discounting the frankly absurd stories about flying saucers and ancient astronauts, we are bound to say that there is no evidence that this has ever happened, though it certainly cannot be ruled out in the future. Meanwhile, we have to resign ourselves to the fact that if we are to detect other civilizations, it can only be by radio signals, which of course travel at the same speed as light.

We have already reached the Moon, and unmanned probes have bypassed many of the planets; by 1990 all the principal members of the Sun's family will have been studied from close range, apart from Pluto, a curious little world, half-rock and half-ice, which seems to be in a class of its own and has been generally relegated to junior status. We have also tried to pick up intelligible radio signals from far-away systems, and we have sent messages of our own. The lack of success is hardly surprising, and in any case we are still limited by the velocity of light. An inhabitant of a planet orbiting, say, the Pole Star would have no hope of hearing us; the Pole Star is nearly 700 light-years away, and a message transmitted this evening would not arrive there until AD 2700 or thereabouts. As we have noted, our own broadcasts began around 1920, so that they have reached only about 60 light-years into space. There are not many sunlike stars within 60 light-years of us, though admittedly there are a few.

Suppose, too, that an alien craft, capable of travelling much faster than light, came down on Earth 1,000,000 years ago, which is not much on the cosmical time scale. The visitor would have found a chilly world, certainly teeming with animal life, but not with anything which could be even remotely regarded as a civilization; and he would probably make notes, depart quickly, and report to his superiors that Planet Three of the Solar System contained no beings classifiable as intelligent. Contact involves not only going to the right place but also going there at the right time.

Before the Space Age, which began abruptly on 4 October 1957 with the launching of Sputnik 1, Russia's first artificial satellite, we had to be content with exploring the universe by means of telescopes and other ground-based equipment. By now we have progressed further, but still only within the Solar System. The Space Telescope, due to be sent up in 1985, should allow us to look further into the depths than ever before, because it will be free of the Earth's dirty, unsteady and obscuring atmosphere, but even then it will have marked limitations. It may just be able to show planets moving round some of the nearer stars, but we cannot be sure; we must wait and see.

Meanwhile, there is no reason why we should not make our own journey into space, visiting other worlds in turn and seeing what we can make of them. At first we can use rockets; after that, we must emulate our Andromedan and fasten ourselves to a light-beam. It will be a fascinating trip, and I invite you to come with me. Let us make a start on the journey.

2 The First Few Seconds

In one way, travelling at the speed of light is a marked disadvantage. Though it will take us hours to reach the more distant planets and years to reach even the nearest star beyond the Sun, we will flash past our Moon in only a second and a quarter. This will give us very little opportunity to look at it closely. Remember, the Moon is very near on the cosmical scale; if you fly ten times round the Earth in a jet aircraft, which does not take an inordinately long time, you will cover a distance equal to that which separates us from the Moon.

Moreover, the Moon has been visited, so there is no need to draw upon our imagination. In the early hours of 21 July 1969, Neil Armstrong made his famous 'one small step' on to the barren rocks of the lunar Sea of Tranquillity; by now twelve men altogether have been there, and scientific equipment has been set up in half a dozen different areas. Indeed, it has been said that the Moon has been so littered with material left behind by the Apollo astronauts and the Russian automatic probes that it looks like a picnic site. This is not quite accurate; what has been left amounts to the equipment of the ALSEPs (Apollo Lunar Surface Experimental Packages), while there are also two Soviet Lunokhods, which moved around the lunar surface until their power gave out. There are even three LRVs (Lunar Roving Vehicles), the moon-cars left behind by the last three American teams, those of Apollos 15, 16 and 17. Since the Moon has no atmosphere, there is no reason why they should deteriorate; unlike terrestrial cars they will not rust, and the next visitors should be able to go across to them, install new batteries, and drive happily away.

The Moon may be parochial, but I feel that we must say something about it, even though in our imaginary journey we will leave it behind so quickly. First, it is not like the Earth except inasmuch as it is made of the same materials. Its surface is mountainous and crater-scarred; there are broad dark plains which are still miscalled 'seas' even though there has never been any water in them (in fact, water has never existed on the Moon). The reason for the lack of atmosphere is that the Moon is not only smaller than the Earth but also much less massive. It would take 81 Moons to balance the weight of the Earth, and this means that the lunar gravity is much weaker. The escape velocity is a mere one and a half miles per second; in other words, if you start moving upward at one and a half miles per second you will break free from the Moon altogether. The Earth's escape velocity is seven miles per second (roughly 25,000 miles per hour) and this means that our planet has been able to hold on to the quickly-moving atoms and atom-groups which make up the air you and I breathe. The Moon was too feeble; any atmosphere it may once have had has long since leaked away into space, so that today the Moon has no atmosphere at all.

The Moon. This was the view from Apollo 17 in December 1972. The bright areas, the craters and the waterless seas are splendidly shown, together with some regions which can never be observed from Earth.

Passing over the surface, and slowing ourselves down so as to have a better view, we see that everything is sharp and clear-cut; there are no clouds or mists, and visibility is perfect all the time. Of course, the dark plains are dominant. Most of them make up a connected system on the side of the Moon which faces the Earth, and they give every impression of having been once filled, not with water but with lava. Evidently the Moon was a highly volcanic world earlier in its history. So far as the craters are concerned, opinions are divided. Some astronomers believe them to be of volcanic origin, while others maintain that they were produced several thousands of millions of years ago by a kind of cosmical bombardment, when meteorites battered the still-solidifying crust and left the immense structures which we see today. No doubt

The first actual voyage to the Moon, Apollo 11, in July 1969. Colonel Edwin Aldrin sets up scientific equipment, photographed by Neil Armstrong. The lunar module appears in the background – the astronauts' only link with home.

Opposite: the Lunar Apennines, July 1971. Astronaut David Scott, of Apollo 15, salutes the American flag. In the background, rises the mountain Hadley Delta. Though it is daylight, the sky is black – as the lunar sky always is.

The Moon-car or Lunar Roving Vehicle, taken to the Moon in Apollo 15, and used by Astronauts David Scott and James Irwin to drive around the surface, reaching the very edge of the great chasm known as the Hadley Rill.

both processes have operated, but as we swoop down to have a really close look we can see no trace of activity now. The Moon is to all intents and purposes inert; its active history ended long ago, well before the first civilizations appeared on Earth and even before the age of the dinosaurs.

From a height of a few miles (and we have been forced to slow down to a cosmic crawl) it is possible to see the small craterlets and hummocks described by the astronauts who have walked among them. The comment 'magnificent desolation', made by the second man on the Moon, Colonel Edwin

Aldrin, who followed Neil Armstrong down on to the Sea of Tranquillity from the *Eagle*, the lunar module of Apollo 11, has never been bettered. Because of the lack of air, and because the Moon's surface curves more sharply than that of the Earth, distances are difficult to estimate. A hill that seems only a few hundred yards away may in fact be at a range of several miles. The craters have sunken floors, with walls which rise to only modest heights above the outer surface; some of them have central mountains or mountain groups, and few parts of the Moon are entirely free of them.

By careful manoeuvring, let us shift round to the far side of the Moon, which is always turned away from the Earth. It has long been known that the Moon's rotation period – in other words its 'day' is 27.3 times as long as ours. This is also the time taken for the Moon to complete one orbit round the Earth, and therefore we always see the same hemisphere. Before 1959, when the Russians sent their automatic probe Luna 3 on a round trip and obtained pictures of the unknown regions, we had no real idea of what the far side of the Moon was like, but it turned out to be just as barren, just as cratered and just as lifeless as the side we have always known. One feature stands out: the huge, dark-floored walled plain which was named Tsiolkovskii in honour of the Russian

theorist who was writing intelligently about space travel more than 80 years ago.

In our voyage of imagination we will not pause to land. If we did, one awkward circumstance would become very evident: the penetrative power of the lunar dust. Commander Eugene Cernan, of Apollo 17, described it as being 'the most awkward hazard of all. It's like graphite; but graphite lubricates, whereas lunar dust makes things stick together. It gets into your space suits, and all moving parts of the vehicles. After I'd handled lunar rocks, it took me weeks to get rid of the last traces of dust which had penetrated into my skin.'

If we did land, we would also have to remember that a very effective pressure

Earthrise. As seen from Apollo 17 in December 1972, the crescent Earth appears above the horizon of the desolate lunar landscape.

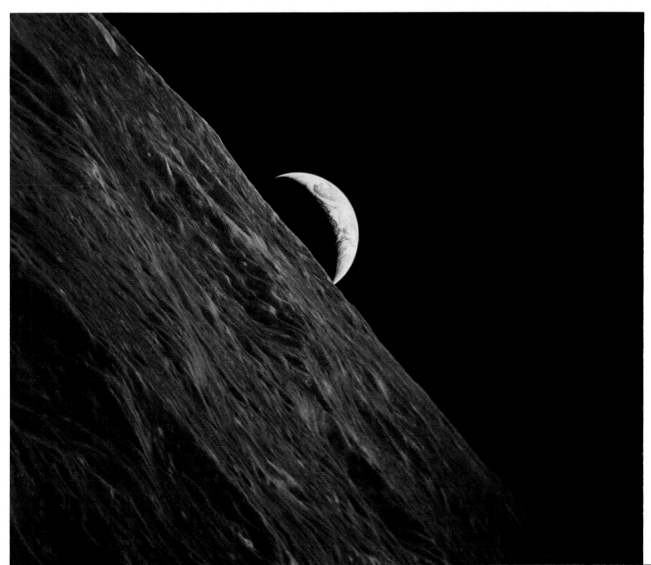

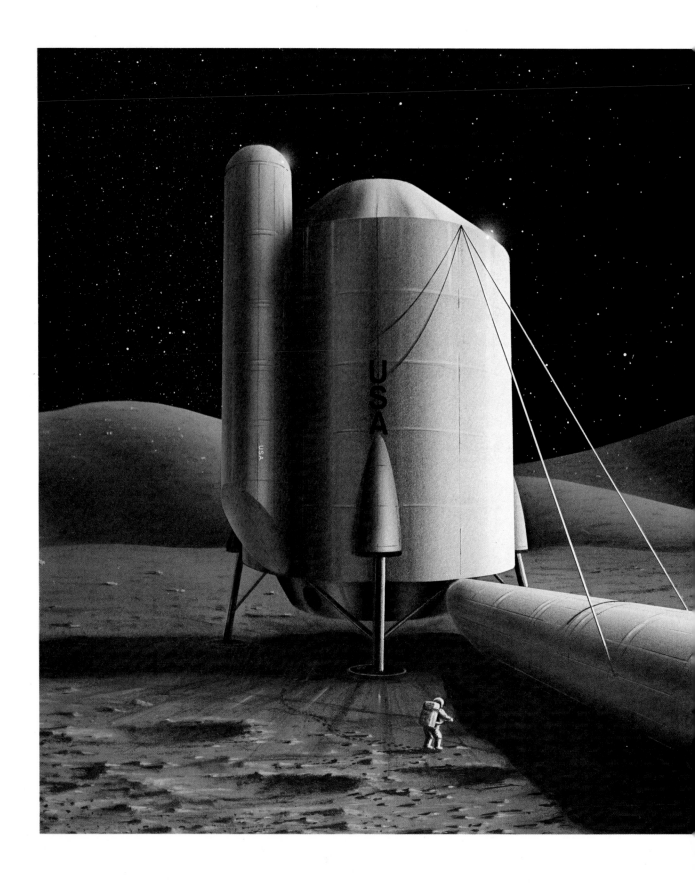

suit is needed. Not only is there a lack of air to breathe but there is also a total lack of pressure, and an unprotected human body would simply burst. Certainly the view from the lunar surface is spectacular. The Sun shines down from a black sky, and from the Earth-facing hemisphere there is also our own world, blue, beautiful and in part cloud-hidden, so that it is not easy to distinguish the outlines of the seas and continents. The low gravity on the Moon means that an astronaut has only one-sixth of his normal weight, and this makes walking around unfamiliar; as Cernan said, 'If you want to change your direction, you have to make up your mind several steps beforehand', as otherwise you will be almost certain to topple over.

Looking down on the Moon from our vantage point, we must ask ourselves about the chances of setting up a manned base there. In fact the chances are very good indeed. From a purely scientific point of view a base could certainly be established within the next few decades, to the enormous benefit of all humanity. Whether or not this will happen depends, sadly, not upon science but upon politics and finance.

The Moon is a fascinating world, and it has lost none of its magic even though we can now handle samples from it in our laboratories. But to go there, there is no need to travel at the speed of light; the first Apollo mission there and back was completed in only eight days. So let us move on, leaving the Earth-Moon system behind, and accelerate again to the speed of light. Within two and a half minutes we will arrive at the first planet: Venus, the Earth's non-identical twin.

Impression of a Lunar base. Lunar bases have been planned, and a base, inside which astronauts can live and work, may look very much like this. The Earth shines down from the blackness of the Lunar sky.

3 Within the Hour

Travelling by rocket, it is essential to do everything possible to save fuel. A rocket works by what Sir Isaac Newton called the principle of reaction: every action has an equal and opposite reaction. With a November the Fifth firework, the fuel is gunpowder; when you 'light the blue touchpaper and retire immediately' the powder burns, forcing gas out through the exhaust and propelling the tube of the rocket in the opposite direction. A spacecraft works in the same way, though the gunpowder is replaced by an immensely complicated motor involving thousands of moving parts. The great advantage is that a rocket pushes against itself, so to speak, and does not depend upon the presence of air around it; it is at its best in the vacuum of space. Unfortunately, no fuels are powerful enough to satisfy the designers. To reach a planet such as Venus, which moves round the Sun rather than the Earth, the only possibility is to use power for the launch and then put the vehicle into what is called a transfer orbit, so that it coasts inwards towards the Sun and meets its target planet at a predetermined point. Where Venus is concerned this takes several months, which is one reason why no manned expedition can be made there yet.

Moving at the speed of light, we can leap across the gap from Earth to Venus in only a little over two minutes, provided that we choose a time when the two worlds are at their closest together and are separated by less than 25,000,000 miles (roughly 100 times the distance between the Earth and the Moon). Again we must slow ourselves

down, or else Venus, like the Moon, will flash past so quickly that we will have no opportunity to see what it is like. Moreover, the view from even a few hundreds of miles is not encouraging. Venus shows us no solid surface; instead, all we can make out is the top of a dense cloud layer. It is always cloudy on Venus, which is why the planet appears so brilliant as seen from Earth; clouds reflect the sunlight very well, far better than the dull, yellowish-grey rocks of the Moon.

Venus is so like the Earth in size and mass that if they were represented by snooker balls, it would be difficult to tell which was which. But there the resemblance ends. Since Venus is on average only 67,000,000 miles from the Sun, as against our 93,000,000 miles, we would expect it to be hotter, but we also have to reckon with the dense atmosphere, which is made up not of oxygen and nitrogen, as ours is, but of the unbreathable gas carbon dioxide. Carbon dioxide acts in the manner of a greenhouse, and shuts in the Sun's heat, so that the temperature below the cloud-deck is very high indeed.

Hovering above the clouds, we can see that they are moving in a curious manner. We know that the solid body of the planet takes 243 days to spin once round, which is longer than Venus' 'year' of 224.7 Earth-days, but the uppermost clouds need only

Impression of Icarus; the strange little asteroid which passes closer to the Sun than the orbit of Mercury, so that when near perihelion it must be red-hot – though when furthest from the Sun it becomes bitterly cold.

four days. To make things even more remarkable, Venus spins from east to west, not west to east in the same sense as the Earth and most of the other planets. If you could see the Sun from the surface of Venus, it would rise in the west, cross the sky very slowly, and set in the east after a lapse of 118 Earth-days.

But could one see the Sun from below those clouds? Russian and American automatic rockets that have landed there show that anything of the sort is out of the question. The Sun would never be seen, and its presence would be betrayed only by a dull glare in the bright orange sky. So let us do what no man has yet done, or will do in the foreseeable future – go and look.

The top of the atmosphere is about 250 miles above ground level. Nothing can be seen below except a dense haze, and even from 125 miles the appearance is much the same. Dropping to 60 miles, we find that the temperature is still far below freezing-point; but as we continue to fall, the atmosphere becomes denser and the temperature rises. At 45 miles we enter the first real clouds, though the visibility is not so restricted as might have been expected. At 40 miles the Sun can no longer be seen clearly, and is nothing more than a diffuse glare. There are liquid droplets in the clouds, and we find that the atmospheric pressure is about half that of the Earth's air at sea-level.

Rain? Unfortunately, no. The droplets in the clouds are not water; they are deadly sulphuric acid, which can attack materials making up ordinary rockets. As we drop lower and lower, the temperature increases alarmingly. At 30 miles our thermometer registers 450 degrees Fahrenheit, over twice that of boiling water. Here, too, the clouds are comparatively dense, and visibility is limited to a few miles.

Lower down still, visibility improves again, and we pass through the bottom of the cloud-deck at about twenty miles above the

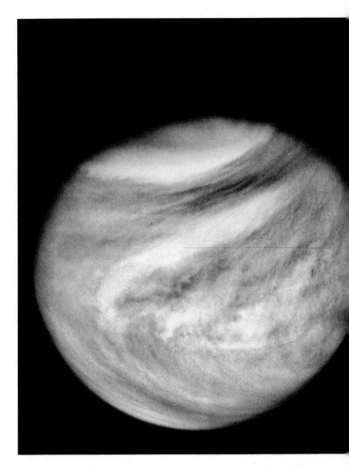

Venus from Pioneer 10. As we close in, we are able to see the cloudy streaks across Venus that run parallel with the planet's equator. The clouds are constantly changing, and the patterns alter quickly.

ground. The 'rain' has ceased; the sulphuric acid particles have been evaporated by the intense heat, which is now over 300 degrees. The outside light is orange-red, and we have entered a region which may be called super-heated, corrosive smog. At a height of five miles the ground below comes into view at last, and the surface features are visible; at last we touch down, to find ourselves in a furnace-like environment with a temperature of nearly 1,000 degrees Fahrenheit and an atmospheric pressure of between 90 and 100 times that on the surface of the Earth.

Incidentally, this pressure caused great trouble with the first rockets that were sent to land on Venus. They were Russian-built, and every precaution had been taken – or so the planners believed. The rockets were even chilled before plunging into the hot clouds. Yet they stopped transmitting well before they were expected to, and it was later found that they had been literally squashed by the tremendous pressure. The later Soviet probes were toughened, but even so they were able to transmit for only an hour or two after arrival before being put permanently out of action.

On Venus, the general scene is very much like the conventional picture of hell. The sky is brilliant orange, while the ground is so hot that it actually glows. The winds are sluggish, at only a few knots, but in that thick atmosphere they have tremendous force, and the various probes that have landed there will certainly not survive for long, unlike those which have come down upon the inert surface of the Moon. If we could

The surface of Venus. Using our special equipment, we penetrate the atmosphere of Venus and see the actual surface. The colours indicate heights rather than true colours; blue indicates low-lying land; green, medium, and yellow, high.

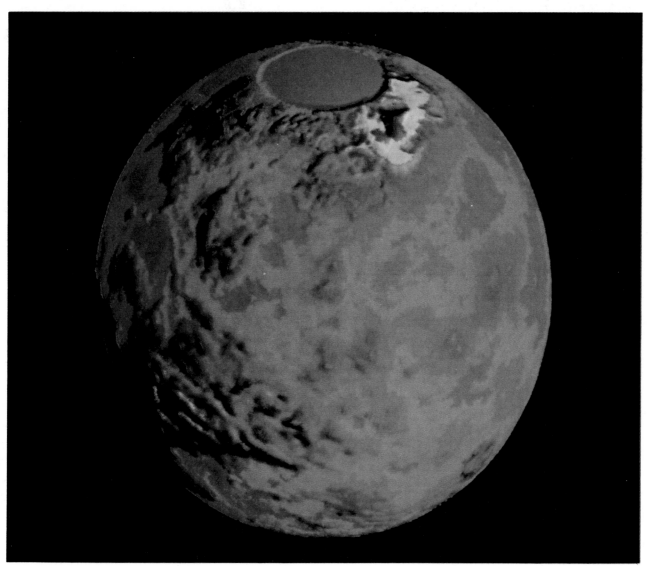

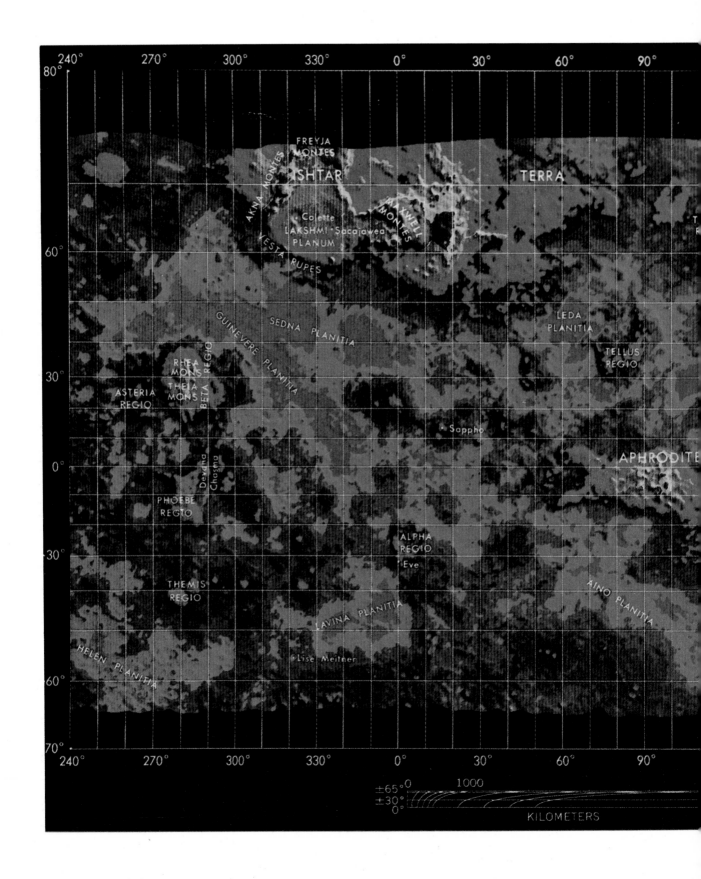

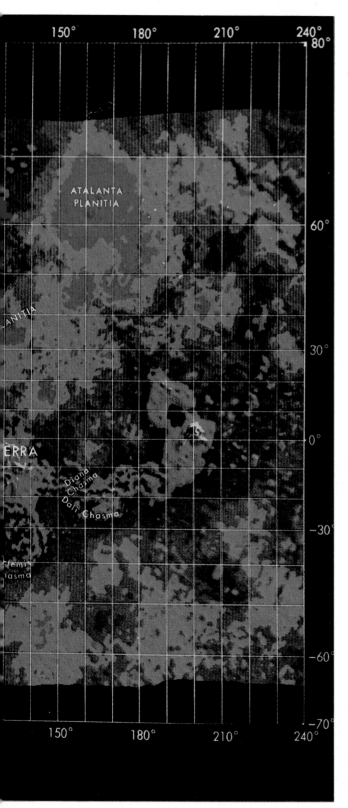

Mapping Venus. There are two main upland areas, Ishtar and Aphrodite (green) and high mountains (the Maxwell Mountains, red); a smaller upland area contains two active volcanoes, Rhea and Theia (left). The lower-lying land is shown as blue.

venture as far as the poles of Venus we would be little better off; the temperature there is much the same as it is at the equator.

The surface is not uniform. Much of it is covered by a huge, rolling plain, as has been known ever since an American Pioneer probe was put into a path above the clouds and mapped the surface by using radar, which, unlike light, can penetrate Venus' protective shroud. There are two highlands, Aphrodite near the equator and Ishtar in the north; if we equip ourselves with 'radar eyes' and hover above Ishtar we will see a smooth central area with huge escarpments around, while the adjacent Maxwell Mountains rise higher than our Everest, and there is one huge crater 3,000 feet deep. Aphrodite is even larger; to its east end is an immense rift valley 1,400 miles long and as much as 175 miles wide in places. More fascinating still is the smaller upland area known as Beta Regio, where we find two massive volcanoes, Rhea and Theia. They are of exactly the same type as the shield volcanoes in our own Hawaiian Islands, Mauna Loa (which is active) and Mauna Kea (which is extinct – at least, one hopes so, because one of the world's major observatories has been built on top of it). There is every reason to believe that Rhea and Theia, too, are active. To complete the picture, there is almost continuous lightning, accompanied by peals of thunder.

Why is Venus so unlike the Earth? The reason must be found mainly in its lesser distance from the Sun. If we could travel backwards in time, so as to see the Solar System as it used to be around 4,500 million years ago, we might find that Venus and the Earth were evolving along the same lines, with similar atmospheres and plenty of surface water. But – and this is the vital fact –

the Sun was roughly 30 per cent less luminous than it is now. As the Sun's power steadily grew, the Earth was sufficiently far out to avoid the worst of the effects; Venus was not. The oceans of Venus boiled and evaporated; the carbonates were driven out of the rocks into the atmosphere, and as Venus turned into the inferno that it is today, all life there was ruthlessly wiped out. This may not have been precisely what happened, but many astronomers now believe so, in which case Venus is a tragic world. It is sobering to reflect that if the Earth had been a mere 20,000,000 miles closer to the

Sun, it, too, would have become a furnace, and you and I would never have existed.

Perhaps we have seen as much of Venus as we want to do on our journey. So we will leave, probably with regret that the planet named in honour of the Goddess of Love has turned out to be so unloving. What will be our next target?

Another two and a half minutes, flying at the speed of light, will take us to the innermost planet, Mercury. Here, too, we know what to expect, because in 1974 an American unmanned probe, Mariner 10, bypassed it after having made a close approach to Venus; in fact, Venus' gravitational pull was used to throw the Mariner into its path towards Mercury – a principle which is sometimes irreverently termed

Impression of the desert of Venus. Venus has no sea; it is bone-dry, and much of the surface is occupied by a vast, rolling plain which is more unfriendly than any Earth desert. We can see volcanoes in the background.

By choosing the right path, we can come down to see Venera 13, the old Russian probe which landed on Venus in 1982 and sent back pictures direct from the surface.

interplanetary snooker. Mariner 10 made three active passes of Mercury before its batteries failed, and sent back excellent pictures of almost half the total surface of the planet.

At first sight Mercury, from close range, appears to be strikingly like the Moon. Here, too, there are mountains, valleys, hills, cracks and craters, though fewer of the dark 'seas'. There is practically no atmosphere, so that it will be safe to land, assuming that we can tolerate the very high temperature. Oddly enough, Mercury is not as hot as Venus. It is closer to the Sun (36,000,000 miles on average), but there is no blanket of atmosphere to produce a greenhouse effect, and the bleak surface is exposed to the full fury of the Sun's rays. The light is blinding; the sky is black, and the Sun is swollen and menacing.

Science-fiction writers used to make use of what was called Mercury's synchronous rotation. It was believed that the rotation period was the same as the revolution period round the Sun – 88 Earth-days – in which case Mercury would keep the same face turned permanently Sunward, just as the Moon does with respect to the Earth. There would be an area of everlasting day and an area of perpetual night, between which there would be a 'twilight zone' from which the Sun would bob up and down over the horizon. Here, it was said, the temperature would be bearable, and landings might be made.

Unfortunately, it has now been found that this whole picture is wrong. Mercury spins round not in 88 days but in $58\frac{1}{2}$ days, so that all parts of the surface are in sunlight at various times, and there is no twilight zone, which makes the planet even more unfriendly than it would otherwise be.

As we move across the surface we come to a series of craters, valleys and mountains. There is also one huge ringed plain, the Caloris Basin, which looks a little like one of the waterless sea-basins on the Moon. If we are prepared to linger there, we will observe some very curious phenomena.

Mercury's orbit is not circular; it is markedly elliptical, so that its distance from the Sun is not always the same. It ranges between 29,000,000 miles at closest approach (perihelion) to 43,000,000 miles at furthest recession (aphelion). When Mercury is at perihelion, and therefore receiving the maximum amount of heat, the Sun is overhead at the Caloris Basin. Just before reaching the exact zenith, or over-head point, it will slow down; after passing across the zenith it will stop and backtrack for eight Earth-days before continuing in the original direction. As it drops to the horizon it will shrink, finally setting 88 Earth-days after having risen. After a further 88 days it will reappear over the eastern horizon, and the whole sequence will be repeated.

There is no mystery about this. Obeying the traffic laws of the Solar System, Mercury moves at its fastest when nearest the Sun, so that at perihelion the angular orbital velo-city is greater than the constant spin velocity of the globe – hence the Sun's habit of backtracking. If we move ourselves to a point on Mercury's surface 90 degrees away from Caloris, things will be different. This time the Sun will rise when the planet is near perihelion, but it will not rise steadily; it will come into view and then sink again, almost disappearing before it rises once more. There will be no hovering near the zenith, but at sunset the brilliant disk will disap-pear, rise again briefly as though bidding adieu, and then finally depart, not to rise again for another 88 Earth-days.

All this is very interesting, but once again we have come to a world which is so inhospitable that we cannot hope to find any form of life there. The craters of Mercury are as sterile as those of the Moon. Possibly an automatic recording station will be

Eruption! Impression of Rhea Mons, one of Venus' active volcanoes, erupting in a blaze of fire, while the lightning flashes and there is continuous thunder. The orange colour of the sky is temporarily hidden by the material hurled out from the volcano.

dispatched there within the next few years, and Mercury would certainly be a splendid vantage point from which to study the Sun, but in its own peculiar way it is almost as hostile as Venus and certainly worse than the Moon.

No planet exists closer in than Mercury; but it would be wrong to claim that the region very near the Sun is empty apart from the thinly spread material that pervades the whole of the Solar System. If we return to our light-beam, we could reach the Sun in a further four minutes, but there might well be other interesting objects on view. One of these is the tiny asteroid Icarus.

The craters of Mercury. Mercury has almost no atmosphere, so that we can see through to the surface. Much of the visible region is highland, and there are many craters, reminding us of the rough upland areas of the Moon.

Asteroids, as we will find when we leave the inner Solar System, are dwarf worlds. Few of them are more than 20 or 30 miles in diameter, and only one, Ceres, has a diameter of as much as 600 miles. Most of the rest are true midgets, and among these there are some that break free from the main swarm, which lies between the orbit of Mars (the outermost of the inner planets) and that of Jupiter (the innermost of the giants). Icarus is one. Its very eccentric orbit takes it from well beyond Mars in to a distance of only 17,000,000 miles from the Sun. It takes 409 days to complete one orbit. When near aphelion it is bitterly cold; when near perihelion it must be red hot. It is only about a mile across, and if we swoop down we will see that it is irregular in shape, probably pitted with craters. Surely Icarus must have

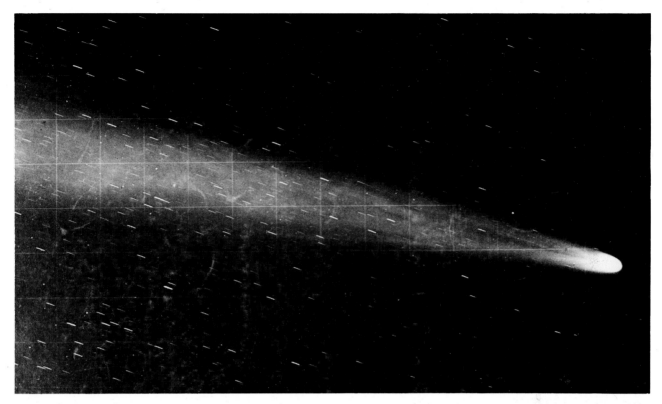

Halley's Comet, a wanderer in space, returning to the Sun every 76 years, but now within our range. As we approach the comet before preparing to plunge into its head, we can see the long tail streaming outwards, pointing away from the Sun.

the most uncomfortable climate of all the planets. We may well regard it as the Devil's Island of the Solar System.

We may also come across an occasional comet in these torrid regions. Comets are not solid, rocky bodies; they are made up of very thin gas together with small, icy particles. Most of them have very elongated orbits, and may take hundreds, thousands or even millions of years to make one circuit, though others have shorter periods of a few years and are familiar members of the Solar System. A great comet may develop a long tail, produced by gases evaporated from the ices in the 'head' together with dust blown outwards by what is termed solar wind – streams of low-energy particles being sent out by the Sun in all directions.

If we use our immunity and approach the Sun, where the temperature is high enough to melt the toughest materials we can devise, we may see a cosmic suicide – a comet actually falling into the Sun. Such events have been recorded. One comet was seen to die in this way in 1979 (though to be more accurate the event happened in 1979; the relevant photographs, taken from an artificial satellite, were not examined until later), and two more have been recorded since. If we are lucky enough to arrive on the scene at the critical moment, we will see the luckless comet rushing inwards, gathering speed all the time, until it strikes the denser part of the Sun's atmosphere at a velocity of up to 350 miles per second. It causes no measurable disturbance in the Sun, because its mass is too small; it simply ceases to exist. How many comets meet their fate in this way is not known, but certainly some of them do, while others break up and leave nothing behind apart from small particles, which we see as shooting-stars if they dash into the Earth's air.

So far, in our voyage of exploration, we have covered considerable distances, but in very quick time. Now let us leave the Sun behind, and move outwards again, working up to our full velocity of 186,000 miles per second. Mercury is bypassed; less than three minutes later, Venus; another three minutes takes us past the Earth and its Moon. Our next target would normally be Mars, the first planet in the Solar System beyond the

orbit of the Earth, but again we may pause at the sight of a comet, either a small one with no tail or else a much larger one with several tails, some made of dust and others of gas. And if we time our journey for the early part of 1986, we can rendezvous with one of the more spectacular members of the comet family: Halley.

Halley's Comet is bright. It is easily visible with the naked eye from Earth when at its best, and it comes back every 76 years, so that we always know when and where to expect it. Its last return was in 1910. The next will be in February 1986 (it was first sighted in October 1982), and there are plans to send five automatic probes to it: two from

An oblique view along the Tharsis Ridge. The three Tharsis volcanoes, Arsia Mons, Pavonis Mons and Ascræus Mons (from bottom) rise to an average height of 10.8 miles above the mean surface level of Mars.

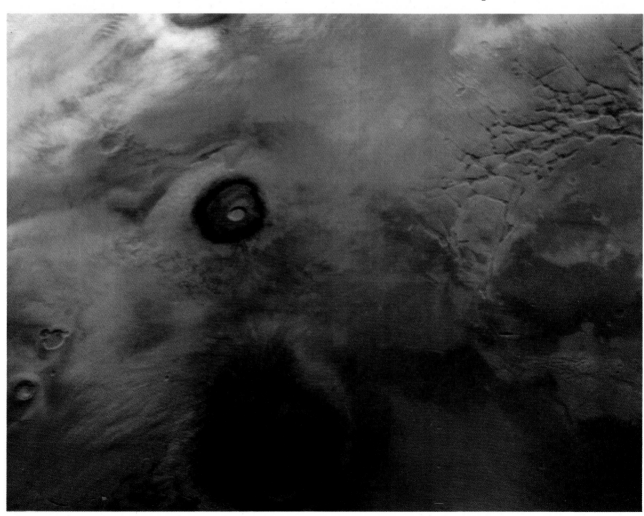

Russia, two from Japan and one from the European Space Agency. What will these probes find?

The trouble is that we have never yet seen a comet's nucleus, the real centre of the comet and the only part which has appreciable mass. This is because every time a comet nears the Sun, its evaporating ices produce a cloud which hides the nucleus completely, and does not dissipate until the comet has moved too far away to be properly studied. Halley provides us with a golden opportunity, because it is the only large comet which we can predict. The various rocket probes will have to reach it by using the clumsy transfer-orbit procedure. On our light-beam, we can do much better.

As we approach, we see the comet clearly – already past perihelion and moving outwards; it is travelling tail first, as all comets do when they have started to recede from the Sun. We will meet it head-on. The comet's filmy mass seems to cover the whole sky; there is nothing definite to tell us when we are inside, but we notice that we are being bombarded by small particles, which may, and probably will, destroy all the planned rockets before they can emerge from the comet's head. The nucleus looks like a large lump of 'dirty ice', a mile or two across, obviously suffering from the effects of evaporation. In a second or two we are through, and Halley's Comet is behind us, still moving outwards. It will continue to do so for another 38 years, by which time it will have penetrated well beyond the orbit of Neptune, the outermost giant planet. Then it will start to travel inwards once more, developing a new tail before it next reaches perihelion in AD 2062. Incidentally, it has retrograde motion: that is to say, it moves round the Sun in a direction opposite to that of the Earth and the other planets.

During our journey to Mars we may well pass other comets, most of them small and tailless. One comet, Encke's, has a period of only 3.3 years, so that it is a regular visitor. But comets are the wraiths of the Solar System, and it is time to turn our attention to the last of the inner planets, Mars.

Again we know very much what to expect. The first rocket to bypass the Red Planet did so as long ago as 1965; eleven years later two American Viking vehicles made gentle landings there and sent back pictures direct from the surface. They even scooped up material from the Martian ground, analysed it, and transmitted the results back to scientists waiting eagerly at home. The main object was to search for life. Disappointingly, no certain trace of any living thing was found, and most astronomers are now coming to the conclusion that Mars is sterile.

It is as yet too early to be certain, but at least we have come a long way since the time of Professor Percival Lowell, a famous American astronomer, who set up a major observatory in Arizona specially to observe the planet, and was convinced that he had seen canals on its surface, built by the local inhabitants to form a planet-wide irrigation system. Lowell knew that the white caps covering the Martian poles are icy, and he believed that an advanced civilization had found a way to use every possible drop of the precious water remaining upon a dried-up world.

Lowell died in 1916. Other observers failed to see the canals, and the rocket probes launched since 1965 have proved that they were due to nothing more than tricks of the eye. But it was a surprise to find that the dark areas, seen clearly on Mars through adequate telescopes whenever the planet is well placed, were not old sea-beds filled with primitive vegetation, as had been widely supposed. The dark areas seem to be the same as the ochre-red tracts, except in colour. Neither are they sea-beds; the most prominent of them, known as the Syrtis Major, is a lofty plateau.

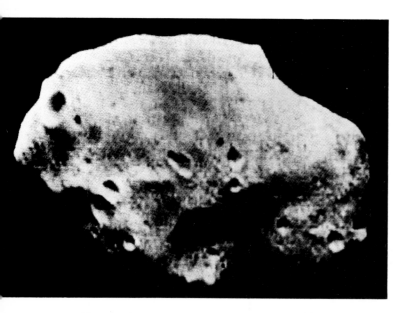

Phobos, the larger and closer-in of the two satellites of Mars, has a dark surface, and neither it nor the other moon, Deimos, would give much help to a Martian observer at night.

Before reaching Mars (which we can do in a little over three minutes after bypassing the Earth, assuming that we choose a moment when the two worlds are closest together), we will notice something else: two specks of light close to the planet. As we draw in, we see that they are irregularly shaped lumps of darkish material, heavily cratered. These are the two moons of Mars, Phobos and Deimos, both discovered by the astronomer Asaph Hall in 1877.

It is worth slowing down to take a closer look at these midgets. Deimos, the smaller, has a longest diameter of less than twenty miles; there are plenty of craters, none of them very deep, but no high peaks. 'Landing' would be more in the nature of a 'docking' manœuvre, as has been carried out many times with artificial satellites circling the Earth. The gravitational pull of Deimos is so weak that a visitor would have practically no weight. To jump up would mean soaring far away from the satellite, and it would be easy to throw a cricket ball clear of Deimos altogether, because the escape velocity is so low.

Mars is superb in the sky of Deimos, but the view from the inner and slightly larger satellite, Phobos, is even more spectacular. Mars dominates the whole scene: sometimes full, sometimes half or crescent and sometimes new. When the Sun is behind it as seen from Phobos, the planet produces a huge, black void, which blots out the light of stars beyond. If we stay awhile, we will notice something else. Mars has a 'day' about half an hour longer than our own (24 hours 37½ minutes), but Phobos has a revolution period of only 7 hours 39 minutes, so that it actually outpaces the spinning planet below. Seen from Mars, Phobos would rise in the west, cross the sky 'the wrong way' and set in the east only four and a half hours later, repeating the cycle three times every Martian day or 'sol'. Yet both it and Deimos are so small that they are of little use as providers of light during the Martian nights. Indeed, Deimos would look like nothing more than a rather large star. There are no seas on Mars; but even if there were, Phobos and Deimos would be unable to produce any appreciable tides.

Rockets that have touched down on Mars have had to use a combination of parachute braking and engine braking. Parachutes are of limited use, because the atmosphere is so thin; it is much less dense than the Earth's air at the top of Everest, and the surface pressure is below ten millibars everywhere, as against a value of around 1,000 millibars for the Earth's air at sea-level. Also, the Martian atmosphere is made up chiefly of carbon dioxide, though it is much too tenuous to cause any Venus-like greenhouse effect.

During the descent to Mars, we cannot help becoming aware that the surface is very far from being the same everywhere. The dark V-shaped Syrtis Major shows up clearly, but its edges are not nearly so sharp and clear-cut as they appear in Earth-based telescopes. In the north there is a prominent wedge-shaped dark region, Acidalia. But our main attention is focused upon a string of huge shield volcanoes spread out along the ridge, which has been named Tharsis. The loftiest of them all is aptly called Olympus Mons or Mount Olympus; it towers to fifteen miles above the ground below, and at its summit there is a 40-mile volcanic crater of the type known as a caldera. Other volcanoes in the Tharsis area are hardly inferior, and there are volcanoes, too, elsewhere on the planet. Once again we

have a cratered scene. The craters are not identical in form with those of the Moon or Mercury, and there are 'chaotic' regions which are almost free from them, but they exist in their thousands.

Coming down still lower, we start to see features that look like old stream-beds; so like, in fact, that it is almost impossible to doubt that they really were cut by running water. There are systems of canyons that dwarf our Grand Canyon in the Colorado, and there is one tremendous valley complex, the Valles Marineris, which lies not far from the Tharsis volcanoes and which can be traced for a total length of over 2,400 miles. Its maximum width is 125 miles, and it

Mount Olympus. We pass over the highest volcano in the Solar System: Olympus Mons (Mount Olympus) on Mars, with its 375-mile base and its height of 15 miles; at its summit there is a 40-mile caldera. Is Mount Olympus extinct, or merely dormant?

is nearly four miles deep in places, so that it is a truly stunning sight.

The Viking rockets – two of them, identical twins – landed gently, one in the ochre tract of Chryse and the other in a rather similar tract, Utopia. Fortunately they avoided any of the boulder-sized rocks which are strewn all over Mars. If we too come down, we will notice that the day-time sky is pink. The Earth's sky is blue because our air is dense enough to scatter the blue part of sunlight around; the lunar and Mercurian skies are jet-black, and that of Venus bright orange, because the cloud-deck reflects the colour of the surface. With Mars, there is not enough atmosphere to cause blueness, but there is a great deal of fine dust in suspension, and it is this which causes the pinkness, just as the Sun appears red to us when it is low over the horizon. Martian temperatures are low. At noon on the equator, at midsummer, a thermometer would register as much as 50 degrees Fahrenheit, but the nights are far colder than anything we experience on Earth, because the thin Martian atmosphere is so inefficient at shutting in the solar warmth.

With its low pressure, the atmosphere of Mars is more or less useless to us. There can be no hope of going out into the open with no equipment other than a breathing mask, as was once thought possible. The lack of pressure would be fatal, and so full space suits must be worn all the time. In this respect, Mars is no better than the Moon.

Picking up the reddish surface material, we are reminded of rust – and this is not a bad analogy; Mars is a rusty place, and the dark areas seem to be simply areas from which the dust has been removed. There are winds on Mars, and there are dust-storms, some of which can spread over the entire planet and hide the surface features for a few days or weeks. But does Mars possess the one essential ingredient which the Moon lacks: water?

Ice on Mars. There is a thin coating of water ice and dust on the rocks, left after the warmth of the sun has evaporated the solid carbon dioxide in the dust particles of the atmosphere.

The white polar caps give us the answer. They wax and wane with the Martian seasons, which are of the same type as ours, though much longer (the Martian year is equal to 687 Earth-days, or 668 sols). The permanent caps are made up of water ice, though there is also a layer of dry ice, or solid carbon dioxide, deposited during winter. We can see strong evidence of past water activity; there is a little moisture in the atmosphere – in fact, as much as it can hold – and we may be fairly sure that there is ice below the surface. At a slightly greater depth there could even be liquid water.

As we look around at the orange land-scape, with its rocks and craters and indications of old stream-beds, a curious thought comes to mind. The air-pressure on Mars is so low that surface water cannot exist; it would simply evaporate. Therefore, if the stream-beds were once water-filled, there must then have been much more atmosphere around Mars than there is today. Moreover, there does not seem to be much erosion – the wearing away of features by natural causes, as happens so rapidly on Earth. Either Mars lost all the main part of its atmosphere early on, so that the craters and other features are thousands of millions of years old, or else there was a period, only a few tens of thousands of years ago, when the atmosphere was thick enough to allow rain to fall and rivers to flow.

The problem is there, and it has yet to be solved. There have been suggestions that the great Tharsis volcanoes erupt violently now and then, sending out enough gases (including water vapour) to thicken the atmosphere for a brief period; such conditions could not last, because the escape velocity of Mars is only 3.1 miles per second, and an Earth-type atmosphere would leak away. Or it may be that at times one of the polar caps melts, releasing its ice to form water. This too is possible, but there is as yet no proof.

Standing on the surface of Mars, one would be struck by the quietness. The thin air can carry hardly any sound-waves, and all communication would have to be by means of radio. The Sun is smaller and less brilliant than the Sun we know, and at nightfall the Earth-Moon pair shine down from the western sky. When twilight is over, the stars themselves are glorious, with barely any trace of twinkling. The constellations are the same as those we can see: Orion, the Great Bear, the Southern Cross and the rest.

Another thought strikes us. Will it be possible to set up a permanent colony on Mars – and, if so, what form will it take?

In many ways Mars is not too unlike the Earth. There is water supply in the form of ice, and there is a tolerable pull of gravity, so that on Mars an astronaut will have one-third of his Earth weight. The only real obstacle is the lack of a useful atmosphere. This means that any base must be thoroughly sealed, so that once inside the colonists can wear normal clothes and dispense with their cumbersome space garments. One scheme involves a system of domes, kept inflated by the pressure of air inside them. There is nothing outrageous in this idea, and the technical problems seem by no means insoluble.

Obviously it will not be possible to do quick 'there and back' trips to Mars, as has been done with the Moon. A rocket, even of improved type, must face a journey lasting for weeks, so that when the time comes to send men to Mars there must be a full-scale expedition rather than an Apollo-type reconnaissance. A base must be erected at once, and all materials brought from Earth must be recycled. It is not likely that any edible plants could be made to grow in the open, though no doubt they could be cultivated inside the base itself.

At all events, we must agree that if we are ever to extend our territory beyond the Earth-Moon system, we must go first to Mars. This raises another interesting point. If a colony is set up there, it will contain men, women and children. We cannot be confident that a boy or girl born and brought up on Mars, under one-third gravity, would ever be able to come to Earth. It is only too easy to imagine how you would feel if your weight were suddenly increased by a factor of three; your legs would buckle under you. There is at least a chance that before many centuries have passed by there will be two separate branches of *Homo sapiens*: Earthmen, and our relations who have emigrated to Mars and can never visit the planet of their ancestors, though of course they would be quite happy in space or on the Moon.

We have started our journey, and have found much of interest, but at the speed of light we have been on our way for only a few minutes. The next step takes us further. There is much more to see before the Solar System is finally left behind.

4 Worlds of Many Kinds

Back on our light-beam, we next cross a wide gap in the Solar System beyond the orbit of Mars. No more large planets show up for another twenty minutes, which takes us out to more than 250,000,000 miles from the Sun. However, there are plenty of small worlds. We are in the zone of the minor planets or asteroids, of which the largest, though not the brightest, is Ceres.

It is worth pausing to look more closely at a few of these asteroids. Ceres, over 600 miles across, is a rocky globe. From Earth it shows only a tiny disk, and no details have ever been seen on it, but there is a good chance that if we swoop down over it we will see craters; indeed, craters seem to be the norm on worlds with solid surfaces (even Venus has them, though they are shallow and presumably subject to fairly rapid erosion). Many of the smaller asteroids are irregular. Eros, whose orbit swings it away from the main swarm and sometimes brings it within 20,000,000 miles of the Earth, has been likened to a cosmical sausage, nearly twenty miles long but less than ten miles wide. However, we have long since left Eros behind. It never recedes to more than 165,000,000 miles from the Sun, and it spends only part of its time in the main zone.

Before the Space Age, it was often thought that the asteroid belt would be a real danger to man-made probes. Though less than 3,000 of these worldlets have been observed well enough for their orbits to be calculated, it

Impression of Pluto and Charon; two dimly-lit worlds, both smaller than our Moon, part-rock and part-ice. We are so far out in the Solar System that the Sun is just a brilliant point.

was thought that the total population might exceed 40,000, and a collision between a spacecraft and a lump of material the size of a coffee-pot could have only one result. Neither is it possible to avoid the zone, because the asteroids are not confined to the main plane of the Solar System. Pallas, second in size only to Ceres, has an orbital inclination of as much as 34 degrees. So far, fortunately, our fears have been groundless. Four probes, two Pioneers and two Voyagers, passed through the main zone between 1972 and 1980 without being damaged, so that either we have been exceptionally lucky or else (more probably) there are fewer very small asteroids than was once expected.

It would be easy enough for a rocket to dock with an asteroid, because even Ceres has so weak a gravitational pull that there would be virtually no sensation of weight. What, then, about the sky from Ceres itself? The Sun is still intensely brilliant, but the constellations are somewhat confused because of the presence of other star-like objects, some of which crawl perceptibly against the background. From the middle of the zone (and Ceres is very near the middle) many asteroids will be visible with the naked eye. Occasionally they will collide, though it is notable that asteroids, unlike some comets, move round the Sun in the same direction, and there is no fear that an asteroid will meet another member of the swarm head-on.

Incidentally, why do the asteroids exist at all? The answer may well be that when the

Jupiter, from afar; already we can see the main cloud belts, bright areas, and spots (especially the Great Red Spot). The size of the Earth is shown in comparison.

main planets were being formed, around 4,700 million years ago, from a cloud of dust and gas associated with the young Sun, Jupiter was produced at a fairly early stage and its great mass prevented any large planet from condensing anywhere near it, so that the planet-forming material was left scattered around as what might be termed cosmical debris. At any rate, another few minutes' travel on our light-beam will take us clear of the asteroid zone. A flight-time of just over half an hour will bring us to Jupiter, the giant of the Sun's family. Using a 1983-type rocket, we would have to face a journey lasting for almost two years. Thus Voyager 2, launched on 20 August 1977, did not bypass Jupiter until 9 July 1979.

There is another reason why actual manned flights to Jupiter must be ruled out. Quite apart from the fact that the surface is gaseous (and it would be difficult to bring a rocket down on to nothing more substantial than a gas layer), Jupiter is surrounded by zones of intense radiation which would be fatal to any astronaut unwise enough to venture too close. The first Jupiter probe, Pioneer 10, made its pass at 80,000 miles above the cloud-tops, but even so its instruments were nearly put out of action and the paths of later probes were altered, so that they passed quickly over the region above Jupiter's equator, where the radiation is at its worst. On our light-beam we are not so restricted. A quarter of an hour to go; we are still over 2,000,000 miles from Jupiter, but already the planet has become a magnificent spectacle. Its yellowish, flattened disk is crossed by dark belts and brighter zones, and there too is the Great Red Spot, a huge oval more than 20,000 miles long. It was once thought to be an erupting volcano, but we know better today. The Red Spot is a whirling storm – a phenomenon of Jovian 'weather'.

Jupiter is not alone. Attending it we can now make out four conspicuous dots of light and several much fainter ones. These are Jupiter's satellites. They total sixteen, but the outer eight are very small, and as we come into the Jovian system it is obvious that they are very similar to asteroids. Indeed, all the evidence indicates that they really are ex-asteroids which strayed well outside the main zone and were caught by Jupiter's powerful pull. Remember, Jupiter is big enough to engulf over 1,000 bodies the volume of the Earth, and it has 300 times the Earth's mass.

Again we must slow ourselves down; Jupiter and its satellites are much too fascinating to be bypassed in a few seconds, and at the speed of light it would take us only just over six seconds to leap from Callisto, the outermost of the large satellites, down to Jupiter itself. These large satellites are known collectively as the Galileans, because they were observed during the winter of 1609–10 by the pioneer telescopic astronomer Galileo.

Callisto is an oddity. As we draw in towards it, craters come into view – not thinly spread, but covering almost the whole of the surface. There seems to be virtually no level ground anywhere, though there is one huge ringed plain which was discovered by the Voyager probes and which has been given the romantic name of Valhalla.

Callisto is less massive than our Moon, though it is larger. Its surface is icy, and gives the impression of being absolutely dead. We are looking at what may be the most ancient landscape in the Solar System. At least we are outside the most dangerous part of Jupiter's radiation, so that if astronauts ever go to the Jovian system they will presumably select Callisto as their first and probably their only port of call.

Move in; we come next to Ganymede, which has the distinction of being the largest satellite in the Solar System, with a diameter almost exactly the same as that of the planet Mercury. But Ganymede is not like Mercury. It has craters and ridges, and we can make out a huge darkish area which has been appropriately named Galileo Regio; there are also some strange stripes, unlike anything on Callisto. Both these large satellites are without atmospheres, and both

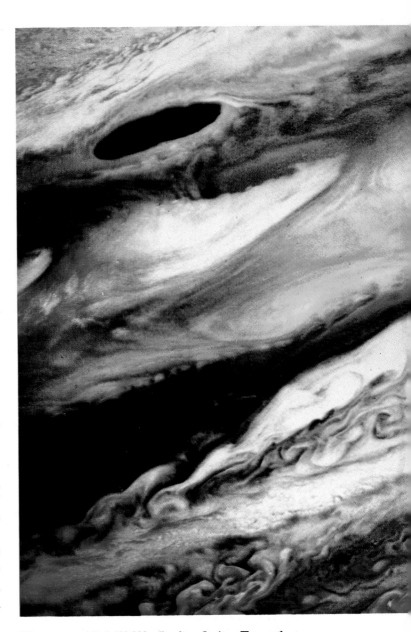

We are now only 6,400,000 miles from Jupiter. The north equatorial belt dominates the scene; a conspicuous dark feature lies close to it. Clouds are in rapid motion; Jupiter is never calm.

are made up of a mixture of rock and ice. For the moment we will not waste time on Ganymede; something much more intriguing lies ahead – Europa.

Here we have a world which is slightly smaller than our Moon. Drawing in towards it we search for the usual craters and hills, but this time the landscape is completely different. Nothing can be made out except for a bright background and a medley of darkish lines which form an intricate maze, calculated to baffle even the most enthusiastic map-maker. Craters are absent; so are any really well-defined features – nothing but these curious lines, and a closer look shows that the lines are very shallow. Europa is as smooth as a snooker ball. The surface is icy, but what lies underneath?

This was discussed in 1982 by two leading American scientists, Steven Squyres and Ray Reynolds, who put forward a theory which sounds like science fiction and yet is

Callisto, Jupiter's fourth satellite, which is made of ice and rock and is larger than our Moon. We can clearly see the great ringed structure which has been named Valhalla, together with many craters and ridges.

based upon sound scientific deduction. As we touch down on Europa we find that the cold is intense, and in places the ice seems rather soft. Burrowing beneath the surface, we pass through a three-mile layer of ice, and then there is a real surprise. Below the ice crust we find an ocean of ordinary water, stretching right round the satellite. It is a deep ocean, and goes down for another 30 miles before giving way to a rocky region which extends down to Europa's core.

There is little light. Occasionally the upper ice layer cracks and water gushes out, freezing as soon as it breaks through the crust and falling back to the surface as frost; each crack may remain open for several years before freezing over again. In the Europan ocean the temperature is still low,

but much warmer than the outer surface, partly because of radioactive materials in the core and partly because Europa is being constantly flexed by the pull of Jupiter, less than 500,000 miles away.

Squyres and Reynolds do not rule out the idea that there may be life in this strange underground sea. After all, living organisms flourish underneath permanent ice in the Earth's Antarctic, where conditions are not a great deal less unpleasant than they may be inside Europa. Of course, any life must be very simple, and the optimistic sportsman who hopes to obtain fishing rights in Europa's sea will be sadly disappointed. Nothing of advanced type could survive there.

The existence of a sub-crustal sea has not been proved. It is no more than a theory, even though a perfectly logical one. At present we can say little more; so let us leave Europa, return to our light-beam and flash in to a rendezvous with the innermost of the Galilean satellites, Io. It will take us less than one second; the minimum distance between Europa and Io is little over 150,000 miles.

Once more the scene is unfamiliar. Ganymede and Callisto are dead, and even the icy surface of Europa shows no signs of activity apart, possibly, from occasional geysers as water from below breaks through the crust. But Io is active, and it is bright red, so that from a distance it looks remarkably like an Italian pizza. There are no craters of the usual type, but there are volcanoes, hurling material high above the red surface. They were discovered by the equipment on board the Voyager probes of 1979, so that at least we are not taken by surprise; they have been given attractive names such as Loki, Pele, Marduk and Prometheus.

Sulphur is everywhere. It has been said that sulphur is Io's equivalent of water, and that the volcanoes are sulphur volcanoes,

Ganymede. As we pass Jupiter's largest satellite, we can see the dark oval Galileo Regio, together with craters and brighter regions. The surface is mainly of ice.

sending blobs of material outwards with a force greater than any volcano on Earth. Landing on Io would be impossible for a manned spaceship, difficult even for an unmanned one; the surface is certainly unstable, and as we come down in our conveniently immune spacecraft of imagination we are conscious that we may have reached what may be the most lethal world in the entire Solar System. Eruptions are constant; here and there we find 'hot spots' with lakes of crusted sulphur, and there is an excessively tenuous atmosphere which would have an evil smell if it were thicker. It is quite possible that the crust overlies an ocean – not of water, like Europa's, but of

Puzzling Europa. We can see no craters or well-defined features on Jupiter's second satellite, only shallow grooves, and it has been said that Europa looks rather like a cracked eggshell.

changed, whereas Callisto, Ganymede and Europa will look just the same millions of years hence.

Leaving Io and continuing on our way to Jupiter, we pass several more very small satellites; one, Amalthea, is red like Io and has at least two peaks, though it seems to be inert, and the redness may have been wafted on to it by material ejected from Io. Then there is Jupiter's ring, quite unlike the magnificent ring-system of Saturn. It is thin, and the particles making it up are dark, which is why the ring has never been seen from Earth and would have remained unknown but for the passing space probes. Next we pass over Jupiter's night hemisphere, the side turned away from the Sun. A brilliant display of aurora stretches for thousands of miles, basically similar in nature to the auroræ or polar lights which we see from high latitudes on Earth and which are caused by electrified particles entering the upper atmosphere. The particles causing our auroræ come from the Sun, though it is more likely that those of the Jovian auroræ are due to the active volcanoes on Io. Over the night side there are lightning flashes, and if we could venture outside our spaceship we would hear incessant thunder. Jupiter is a noisy planet as well as a brilliant one, at least at some wavelengths.

Coming back over the day side, the scene changes again. Jupiter, incidentally, is a quick spinner; its 'day' is less than ten hours long, which is why the globe is so obviously flattened at the poles. Centrifugal force makes the equator bulge out. (This also happens with the Earth, but with us the flattening amounts to only 26 miles as against more than 4,000 miles for Jupiter, partly because the Earth's globe is solid to a great depth and partly because it rotates so much more slowly.) Quite apart from the rapid rotation, Jupiter's surface is in a state of constant turmoil. The Great Red Spot shows

liquid sulphur. To make the situation even worse, a powerful electric current flows between Io and Jupiter, and Io itself moves in the midst of the deadly radiation zone. We are also within Jupiter's magnetic field, much the strongest in the Solar System.

Since the Ionian volcanoes are active today, they must have been erupting for most of the satellite's lifetime; it would be too much to expect that Io would put on a special display for the benefit of our Pioneers and Voyagers! There must be some force which keeps the interior of Io 'stirred', so to speak, and this force must come mainly from Jupiter, though the other Galilean satellites, notably Europa, may be involved to some extent. The liquids below Io's crust are being moved around all the time, so that as one volcano ceases to erupt another begins. The surface of Io is clearly 'young'. In a few years' time the whole aspect will have up vividly, and is seen to be rotating, pulling in and 'squashing' other spots and features that catch it up. It is a high-pressure region, and its top is even colder than the surrounding gaseous surface. According to what we have learned from the spacecraft, the colour of the Great Red Spot is due to phosphine, which gushes up from below and is broken up to make red phosphorus as soon as it meets the sunlight. The Great Red Spot is not unique, but it is much the largest feature of its kind, and it has certainly been in existence for at least 300 years. It sometimes vanishes for a while, but it always returns, though there is evidence that it is shrinking and may not persist for more than another century or two.

Io, world of volcanoes. With its ceaseless activity, its changing sulphur surface and its hot spots, Io is a dangerous world.

Dropping down to the top of the cloud layer, we find that the dark belts are regions where gas is sinking, while the bright zones indicate places where the gas is rising from below. No actual spacecraft has been so close, but one is planned: Project Galileo, due for launching in 1985. Part of the vehicle will stay in orbit round Jupiter, while another section will be sent right into the clouds, continuing to transmit until it is destroyed – as must inevitably happen within an hour after entry.

Let us follow the path which the Galileo probe will take. First we pass through the upper clouds, which are made up of crystals of ammonia; the outside pressure is about the same as that of the Earth's air at sea-level. After passing through a second layer, we come to a region of water-ice crystals.

The pressure is increasing rapidly, and the temperature is rising; Galileo will be dead by the time that it has entered the region below the water-ice clouds, and so we must return to our imaginary vehicle, impervious alike to heat, cold and pressure. Gradually the gas around gives way to liquid, and before long we are in yet another sea, this time made up of hydrogen. Jupiter is liquid for most of the way through to its comparatively small solid core, and it is not surprising that hydrogen is the main constituent. After all, hydrogen is the most plentiful element in the universe.

Deeper still there is another transition layer. The liquid hydrogen changes its nature and begins to assume some of the characteristics of a metal, though this does not make it any the less liquid. We reach a region where the temperature is much the same as on Earth, and there have been some rather far-fetched speculations about Jovian life, though it seems that the conditions are much more forbidding than they are in the hypothetical sea of Europa; hydrogen does not seem to be a promising medium, and any primitive organisms would have to be careful not to swim too high into the cold regions, or dive too low, where the heat is intense. There is no light at all, and by the time we have penetrated to the actual core the temperature has risen to at least 30,000 degrees centigrade. No material probe could withstand this combination of fierce heat and crushing pressure, so that we are limited to informed guesswork.

Jupiter's ring. As we approach Jupiter we can just make out the ring, which is dark and thin, though in this colour composite it appears as two light orange lines protruding outwards to the right from Jupiter's limb.

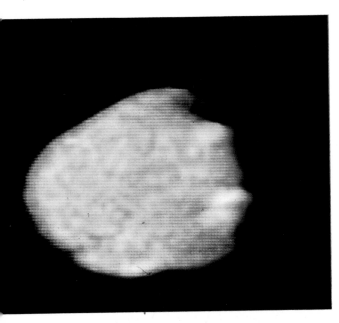

Amalthea, a small, irregular satellite very close to Jupiter. From Earth it is hard to see; as we bypass it, we can make out a reddish surface together with two objects which are probably hills. These are shown to the right.

The internal heat does not mean that Jupiter is a small star. Admittedly it emits more energy than it would do if it depended entirely upon what it receives from the Sun, but probably the excess is nothing more than heat which has been left over, so to speak, from the early days of the Solar System when Jupiter and the other planets were still condensing from the gas-and-dust cloud or solar nebula. Jupiter has simply had insufficient time to cool down. But we have already noted that the Sun and other stars shine by nuclear reactions, which require a temperature of around 10,000,000 degrees, so that Jupiter's 30,000 degrees are not nearly enough. Jupiter is not a star, and can never become one.

Leaving Jupiter we come to another vast gap, this time with no asteroid zone even though we meet with several small bodies; in particular there are two groups of asteroids, the Trojans, which move round the Sun in the same path as Jupiter, though they keep either well ahead of the Giant Planet or else well behind it, and are in no danger of being drawn into the Jovian satellite system. One of these Trojans, Hektor, is either elongated or else double. We do not pass near enough to find out; the most beautiful of all the planets, Saturn, is next on our list.

Looking at a map of the Solar System, it is not easy to realize how spread out the giant planets are. If we leave Earth at the velocity of light and make straight for Jupiter at its closest, we will be just over 30 minutes on the way. We need another 30 minutes to rendezvous with Saturn, so that Jupiter is a half-way house (assuming that the Earth, Jupiter and Saturn are roughly lined up, as does happen regularly; Jupiter takes $11\frac{3}{4}$ years to complete one journey round the Sun, while Saturn takes $29\frac{1}{2}$ years). With Saturn we have some proven facts to guide us, because Pioneer 11 and both the Voyager spacecraft have sent back information from close range. As we move on, Saturn looms larger and larger in the sky; now the rings are visible in all their glory, and Saturn's yellowish disk shows dark belts not unlike those of Jupiter, though the colours are muted and there is nothing so prominent as the Great Red Spot. Saturn, overall, has a somewhat bland appearance, due to overlying 'haze'.

As with Jupiter, so with Saturn; there is a wealth of satellites, and Saturn's total family is now known to contain more than twenty members. But instead of four large satellites and many very small ones, Saturn has a single giant moon, Titan, and several which are of intermediate size, with diameters ranging from 200 to 1,000 miles. The first satellite to be encountered is Phœbe, at a distance of more than 8,000,000 miles from Saturn. Even travelling at the speed of light, it will take us 43 seconds to move from Phœbe to the top of Saturn's cloud layer.

Phœbe is an exceptional body. It is darkish, so that it is presumably rocky rather than icy; it is a dwarf, with an estimated diameter of a mere 99 miles, and it moves round Saturn in a wrong-way or retrograde direction, in the manner of a driver who mistakes his (or her!) way on a roundabout. Phœbe is not unique in this respect, because four of Jupiter's outer satellites also have retrograde motions, but Phœbe is also exceptional inasmuch as it does not have synchronous rotation. It spins round in nine hours, whereas it takes over 550 days to complete one orbit round Saturn. It may well be a captured asteroid, though not many asteroids wander so far from the Sun.

Another half-minute will bring us to the remarkable Iapetus, the zebra of Saturn's family; one part of it is as bright as ice, while another part is blacker than a blackboard. There are craters, and the fact that the overall density of Iapetus is not much greater than that of water seems to show that

The Great Red Spot. We are careful not to go too close to this awe-inspiring feature; a bright red whirling storm, which sucks in or distorts other cloudy features as it rotates. The whole area of the Spot is violently disturbed.

we are seeing a bright satellite covered in part by a dark stain rather than a dark satellite with bright deposits. But what causes the stain? Suggestions that it is material wafted in from Phœbe seem improbable; it is more likely that the dark material has welled out from below the crust, though this must have happened a long time ago, since there is no trace of activity on Iapetus now.

Iapetus is over 1,000,000 miles from Saturn, but it is worth our landing there, because we will have an excellent view of the rings. All the closer satellites move practically in the plane of the ring-system, so that from these satellites the rings will appear edgewise-on as nothing more than a line of light – as sometimes happens when Saturn is observed from Earth, the last occasion being in 1980. The rings are extremely thin, and their thickness is now thought to be less than a mile. The orbit of Iapetus is tilted at an angle of nearly fifteen degrees, so that the rings are well displayed. We can also see the much smaller satellite Hyperion, again with a cratered surface, but irregular in shape; its longest diameter is no more than 225 miles. Hyperion looks almost as though it is part of a larger body which met with disaster. This may be so, but in this case we must admit that there is no trace of the other half!

Then, at only 760,000 miles from Saturn, or less than eight seconds' light-time from Iapetus, we come to Titan, which many astronomers believe to be the most intriguing world in the entire Solar System. From above, nothing can be seen but a uniform layer of cloud. The surface of Titan is hidden as effectively as that of Venus, but the atmosphere is different. Instead of being made up of carbon dioxide, it consists chiefly of nitrogen, the gas which also makes up 78 per cent of the air we breathe on Earth. Moreover, the Titanian atmosphere is thick, giving a surface pressure one and a half

times that of our air at sea-level. There are appreciable quantities of other gases, notably methane (the gas dreaded by miners, who call it fire-damp because under suitable conditions it is dangerously explosive), but nitrogen is dominant.

The Voyagers gave tantalizing pictures of the upper clouds of Titan, but they could do no more, which was doubly unfortunate because the surface conditions are likely to be so extraordinary. The upper atmosphere is layered and orange, and as we plunge into it we come to an extensive hazy region

Hyperion. Our quick fly-by of Hyperion is enough to show that it is irregular, with a longest diameter of only 225 miles. It is unlike the two satellites whose orbits are closest to it – Titan and Iapetus – and may contain more rock than ice.

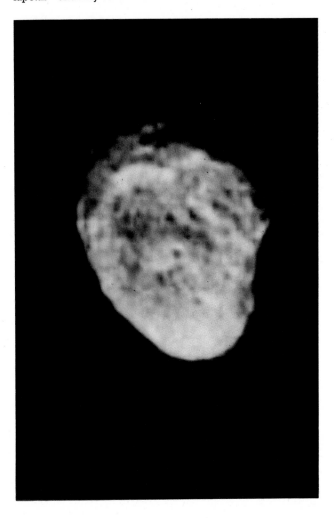

before meeting the denser part of the atmosphere at about 140 miles above the ground. Lower still we reach a cloudy zone, with the clouds made up of methane rather than our familiar water vapour.

The surface is certainly gloomy, since little sunlight can penetrate the clouds. The temperature is low, about −290 degrees Fahrenheit, and this, significantly, is near the triple point of methane – i.e. the temperature at which methane can exist as a gas, liquid or solid, just as, on Earth, H_2O may exist either as water vapour, liquid

Tantalizing Titan, showing only the top of its nitrogen-rich atmosphere. Below there may be a fantastic landscape, with methane cliffs and rivers – possibly even a deep methane sea covering almost the whole of Titan.

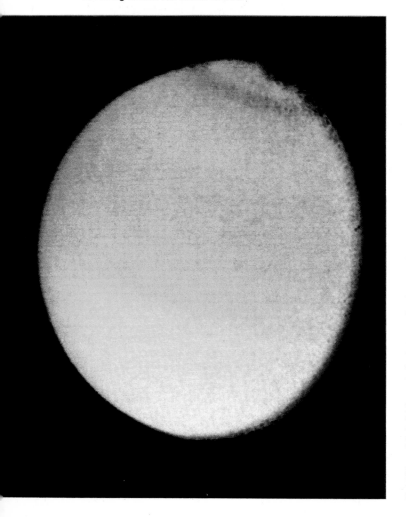

water or ice. On Titan we may well find cliffs of solid methane, rivers of liquid methane, and a steady drip-drip of methane rain from the orange clouds above. There are organic materials in the atmosphere, too, and these may fall constantly; if some of them contain carbon-nitrogen compounds, we may even find deep beds of tar-like material. Titan has a long 'day'; its rotation period is the same length as its revolution period round Saturn (fifteen and a half Earth-days).

A nitrogen atmosphere, organic compounds, possibly oceans of some kind . . . does this show that life could exist on Titan? It has been claimed that Titan is like an Earth in deep freeze, so that the only factor which prevents life from gaining a foothold is the very low temperature. Yet if Titan were warmer, it might lose its atmosphere altogether, because the escape velocity is low. In the far future the Sun will become more luminous as it enters its red giant stage, and Titan's temperature will rise rapidly, but even then the conditions for life do not appear to be promising. Assuming that Titan could hold on to its atmosphere, conditions might become tolerable for a while, but the Sun's red giant stage will be followed by collapse into a very feeble white dwarf condition, and any remaining atmosphere on Titan will freeze into a solid block.

Yet we cannot be sure of anything, and probably we must wait for a special Titan probe to be put into a path round the satellite and map the surface by radar – as has already been done with Venus. Until then we can only speculate, but Titan is in no hurry to give up its secrets. There seems no reason why landings there should not be made eventually. Titan moves almost exactly at the edge of Saturn's main magnetic field, but in any case the Saturnian radiation

Enceladus. Even from close range we can see only small, fresh-looking craters, while some areas are smooth apart from shallow grooves.

zones are much weaker and therefore much less dangerous than those of Jupiter.

The Titanian landscape is unique, and as we rise once more above the orange smog and make our way towards the inner icy satellites we are conscious that we have a great deal to learn yet. Next in line are Rhea, Dione and Tethys, all between 500 and 1,000 miles across and all heavily cratered, though Dione also has strange frost patterns and Rhea's craters give the impression of being very ancient indeed. Tethys is remarkable in having one crater, now named Odysseus, with a diameter one-third that of the satellite itself, plus a long rift, Ithaca

Chasma, which has an average width of 60 miles and a depth of between two and three miles; it runs from the north pole down to the equator and across to the south pole. Tethys appears to be a globe of almost pure ice, and it may well be that the Odysseus crater was produced by internal action, since a meteorite large enough to produce such a crater would probably have shattered Tethys completely.

We are nearing Saturn. We must pause briefly to look at two other satellites, Enceladus, with its highly reflective ice crust and smooth areas, and Mimas, with one huge crater, Herschel, which rivals Odysseus in relative size (though in fact Odysseus is big enough to contain the whole of Mimas). Then there are Janus and Epimetheus, both small and irregularly shaped,

Rhea. With Rhea, we look down upon a landscape which is very heavily cratered; there seems to be no level ground. Rhea gives the impression of being utterly dead. Nothing has happened there for thousands of millions of years.

Dione. We must examine Dione, one of Saturn's satellites. The surface is partly covered with wispy features, possibly fresh frost; in other areas the icy surface is heavily cratered.

which seem to be playing a game of cosmical musical chairs. Their orbits round Saturn are almost the same. Every four years one of them catches up with the other, and the minimum distance between them may be only a few miles; as they draw apart again they actually exchange orbits until the time of the next encounter. There is little doubt that they do represent the two halves of a larger body which has broken up.

By now we are almost at the edge of the ring-system proper. It is an amazing sight. From Earth we can see two bright rings, separated by a gap called the Cassini Division in honour of its seventeenth-century discoverer, and a much fainter one closer to the planet. As Voyager told us, and as we can see as we swoop down, the rings are anything but simple. They consist of small icy particles, but are divided into thousands of separate ringlets and narrow divisions, so that they have more grooves than any gramophone record. Ring F, the first ring outside the main system, is kept firmly in order by two small satellites, one of which is slightly closer in than the ring and the other slightly further out. If a ring particle dares to stray from its proper orbit one or other of the 'shepherd' satellites will force it back again, so that Ring F survives, even though it is kinked in places and is strangely uneven.

As seen from Saturn's surface the rings would form magnificent arcs in the sky, though from some regions they would cover

Saturn. Now we are only 8,000,000 miles from Saturn; the rings are excellently seen, together with two of the medium-sized satellites, Tethys and Dione. Tethys is closer to Saturn and is smaller than Dione.

the Sun for a significant part of Saturn's long year. The surface itself is gaseous, and, as with Jupiter, there are upper clouds and layers, beneath which comes a sea of liquid hydrogen. The solid core is smaller than Jupiter's, though still considerably larger and more massive than the Earth's. The core temperature must exceed 10,000 degrees centigrade, but the cause of the excess radiation is different than in the case of Jupiter, because Saturn is smaller, and has now had enough time to 'cool down' since its formation. What apparently happens is that helium droplets in the upper layers are falling down towards the centre through the lighter hydrogen surrounding them; as they do so, they give off gravitational energy.

Leaving Saturn, we look back once more at the bland globe, the circling satellites and, above all, the complex rings. We do not yet have a full explanation of how the grooves are formed, and neither can we really understand certain peculiar, spoke-like features crossing the brightest ring, known as Ring B. There is still much to be learned, but meanwhile we must prepare for another journey which, even at the speed of light, will take us an hour and a half. We are *en route* for Uranus.

We might have expected the region beyond Saturn to be more or less clear, but now we are coming to something very odd; it looks like an asteroid, darkish in colour, and by asteroidal standards it is large, with a diameter of several hundreds of miles. We have known it since 1977, when it was discovered by the American astronomer Charles Kowal, who named it Chiron after the wise centaur of Greek mythology. Chiron spends its time mainly between the orbits of Saturn and Uranus, and takes 50 years to complete one circuit of the Sun.

Occasionally it can bypass Saturn; in the year 1664 BC it approached Saturn within 10,000,000 miles, so that it was then not far beyond the orbit of Phœbe. Can Chiron be an escaped satellite of Saturn – or is it similar to Phœbe, apart from the fact that Phœbe has been captured? These are problems yet to be solved. At any rate Chiron is a dark, lonely little world, and it may not even be in a stable orbit, so that in the distant future it may be thrown out of the Solar System altogether to become a solitary wanderer in interstellar space.

We have spent a long time in exploring the familiar planets, and we must pass more

Saturn's rings. We are approaching the rings from 'underneath', but they remain visible, as the sunlight can penetrate them. Colours shown here have been enhanced to facilitate analysis.

briefly by the final three; two giants (Uranus and Neptune) and one small world (Pluto). Uranus at its closest is two and a half light-hours away from the Earth, Neptune as much as four light-hours; despite their size – Uranus, with a diameter of over 30,000 miles, is slightly the larger of the two – Earth-based telescopes can show little upon their gaseous surfaces. Uranus, coming first into view, is the only green planet. There are five satellites, all smaller than our Moon and presumably icy and cratered, and there is a

Saturn's red spot. We can still make out a red spot in the upper clouds of Saturn; it was first recorded by Voyager 1 in 1979, and may not be permanent, but we can notice its colour at first glance.

ring-system. Apparently there are at least eight rings, narrow and dark; it has been said that while Saturn's rings are as reflective as snow, those of Uranus are as dark as soot. No Earth telescope will show them clearly, and we must hope for good results from Voyager 2, which is scheduled to pass by Uranus in January 1986.

Surprisingly, Uranus has an axial tilt of more than a right angle, so that at times one of the poles will face the Sun. The calendar is strange by any standards. Each pole has a night lasting for the equivalent of 21 Earth-years, with a corresponding midnight sun at the other pole; subsequently the conditions are reversed, though for part of the 84-year-

long Uranian revolution period the situation is less abnormal.

As we fly past Uranus, we recall something about the story of its discovery. It was found in 1781 by an amateur astronomer named William Herschel, at that time a professional organist and later probably the best observer of all time. Uranus is just visible with the naked eye if you know where to look for it, but Herschel identified it partly because it showed a small disk, which no star can do, and partly because it moved from night to night against the starry background. (At least, Herschel realized that it must be a member of the Solar System; originally he mistook it for a comet.) Later, Uranus was found to be moving in an unexpected manner, and the cause was tracked down to the influence of a more remote planet, now named Neptune. The mathematicians worked out the probable position of Neptune, and when telescopes were turned in that direction the planet was soon discovered. It is too faint to be seen with the naked eye from Earth, but telescopes show a distinct disk, blue instead of green like Uranus.

This time there are no rings, but there is one large satellite, Triton, which may be about the same size as Mercury. It is unique among large satellites inasmuch as it orbits Neptune in a retrograde direction, so that while Neptune rotates from west to east Triton moves from east to west; it may be this which has led to unstable conditions and prevented the formation of a Neptunian ring. At least, we have so far seen no sign of one, though we will not be sure until August 1989, when Voyager 2 is scheduled to bypass Neptune and send back information from close range.

Voyager 2, remember, was launched in 1977. It takes a dozen years to reach Neptune, and even this is only possible because of a fortunate coincidence. Near the end of the 1970s the four giant planets were strung out in a gentle curve – something which happens only once in two centuries – and Voyager 2 was able to use their gravitational fields; Jupiter's force accelerated the Voyager out to Saturn, Saturn sent it on its way to Uranus, and if all goes well Uranus will give it a similar 'push' towards Neptune. Yet a journey of twelve years is very different from the four hours needed for the ray to which we have so conveniently attached ourselves.

At present (1983) Neptune is the most distant known planet in the Sun's family. There remains Pluto, which is a true oddity. It has an eccentric path which brings it within the orbit of Neptune, though there is no danger of a collision on the line, because Pluto's orbit is also tilted at an angle of seventeen degrees. Pluto next reaches its perihelion in 1989; until 1999 its distance from the Sun (and the Earth) will be less than Neptune's. Unfortunately, Voyager 2 cannot be sent anywhere near it. The only way to use the 'slingshot' method would be to dive first several hundred miles below the surface of Neptune, which does not seem very practicable! However, our light-beam can take us there in an hour or two after departing from Neptune, and as we close in we can see that Pluto is quite unlike the other outer planets.

There is not just one Pluto; there are two, making up a double system. One member, Pluto itself, is about 1,800 miles in diameter, which is smaller than our Moon. The other, now named Charon, has a diameter of about 600 miles. As we are nearing the end of our Solar System journey it is worth making a landing, so let us touch down on to Pluto.

We are about 2,770 million miles from the Sun, or more than four light-hours, but the sunlight is still brilliant, casting its radiance over the surface of the icy little world and illuminating it 1,600 times as brightly as the full Moon can light up the Earth. The sky is intensely black. The inner planets are not to

Mimas. Our view of Saturn's famous inner satellite is dominated by the huge crater Herschel, with its high walls and central peak. The scene looks lunar, but there is no real similarity; the Moon is rocky, while Mimas has an icy surface.

be seen, because they are much too close to the Sun, and neither is Jupiter on view; the distance between Jupiter and Pluto is more than five times the distance between Jupiter and the Earth. Saturn may be glimpsed, but even Uranus and Neptune will cut poor figures in the Plutonian sky. To make up for this there is Charon, only 19,000 miles away. It is dim, but still very prominent, and even with the naked eye we can see its surface features. After an hour or so we realize that it has not moved. It revolves round Pluto in six days nine hours. This is the same time as Pluto takes to spin on its axis, so that Charon is permanently fixed, while from the opposite hemisphere of Pluto it can never be seen.

Looking down at our feet, we see that there is frost on Pluto – not ordinary frost, but methane frost. Now that the planet is almost at its closest to the Sun there is a certain amount of evaporation, whereas when Pluto is at its most remote (4,566 million miles, or nearly seven light-hours) the methane becomes solid.

Pluto has set astronomers problem after problem. Like Neptune, it was tracked down mathematically before it was actually seen; the calculations were made by Percival Lowell and the discovery by Clyde Tombaugh, in 1930, at the observatory which Lowell had founded in Arizona. (Sadly, Lowell did not live to see it; he died in 1916.) The trouble is that because of its small size and mass, Pluto is quite incapable of tugging giants such as Uranus and Neptune measurably away from their predicted paths. Either Lowell's prediction was sheer luck, or else Pluto was not the planet for which he had been hunting, and the real culprit still awaits discovery.

Moving outwards from Neptune and Pluto we will have time to ponder on all this, because we have a lengthy journey in front of us. There seems now no doubt that Neptune and Uranus are subject to perturbations by an unknown body, but what can it be? Possibly a new planet, so remote and so faint that it will be excessively hard to locate. Even Pluto is a tiny speck in our best telescopes, and the more distant object, often referred to as Planet X, will be fainter still even if it turns out to be a giant. Trying to identify one minute speck against a background so rich in stars would be a hopeless business without some knowledge of where to look. Uranus and Neptune cannot help us, but there is a chance that we may be able to create the opportunity for ourselves.

Pioneer 10 bypassed Jupiter in December 1973 and then went on its way, so that it will leave the main part of the Solar System in the early 1990s and will never come back. Pioneer 11, which passed Jupiter in December 1974, was then swung back across the planetary system to an encounter with Saturn, after which it too began a never-ending journey. The point is that the two Pioneers are leaving the Solar System in opposite directions, and we are still in radio touch with them. If they show signs of departing from their planned paths, we could have a clue as to the likely position of Planet X.

If one of the Pioneers wanders, but not the other, a planet must be the answer. But if both wander, then the culprit is more likely to be a dead star – a companion of the Sun which has used up all its energy so that it has become dark and invisible. Its distance would have to be of the order of 50,000 million miles, or over 3 light-days. There is nothing improbable in assuming that the Sun might have such a companion; two-star or binary systems are common in the Galaxy, as we will find as we travel on. More exciting still, there could even be a Black Hole at twice this distance. This has been suggested, though it seems that so dramatic an object would have betrayed its presence long before now.

Leaving dead stars and Black Holes out of the reckoning for the moment, we will keep a sharp lookout for Planet X as we draw away from the Sun, accelerating once more to the full velocity of light. Now and again we pass ghostly comets. One of them is Grigg-Mellish, which takes 164 years to complete one orbit and was last at perihelion in 1907, so that it is now almost at its aphelion over 5,000 million miles away; not until 1989 will it start to swing inwards again. But the Solar System is fading; the Sun is no longer a disk but an intensely brilliant point.

We are still in the outer reaches of the Solar System, and after a lapse of between one and two years we come to a vast cloud of comets, all moving sluggishly round the remote Sun. This is, of course, the Oort Cloud. Nobody has ever seen it, and some astronomers even doubt its existence, but if it is real then it must be the source of all the bright and faint comets that come in close enough to be detected from Earth. Some disturbance may pull a comet out of the Oort Cloud and send it plunging Sunwards. Eventually it will reach perihelion, and if it encounters one of the giant planets *en route* it may have its path twisted, so that it becomes a short-period comet. Even Halley, now returning every 76 years, may be a former member of the Oort Cloud.

Now, at last, we are leaving the Sun's realm. Beyond the edge of the Oort Cloud we reach a distance of two light-years from the Sun. Behind us the Sun remains bright, but in the opposite direction there is something even more brilliant: the system of Alpha Centauri. Our exploration of the Galaxy is about to begin.

Impression of Neptune from close range; we pass across one of its satellites, but can see little detail on the blue face of the planet, and there is no sign of a system of rings.

5 The Greater Universe

Once we have cleared the outer boundary of the Oort Cloud we have definitely left the Solar System, and are on our way to Alpha Centauri. Perhaps it would be better to say the 'Alpha Centauri system', because our nearest stellar neighbour is made up of not one star, but three. Two are bright, and comparable with the Sun, while the third, Proxima, is red and feeble. As we speed towards them it is worth casting our thoughts back to the year 1838, when Alpha Centauri became the second star to have its distance announced.

The method, put into practice by a Scottish astronomer named Thomas Henderson, was that of parallax. The best way to demonstrate what this means is to hold up a finger, close one eye, and align your finger with some distant object, such as a tree. Now, without moving your finger or your face, use the other eye. Your finger will no longer be lined up with the tree, because you are observing it from a slightly different direction; your eyes are not in the same place. Measuring the amount of the angular shift (parallax) and knowing the distance between your eyes means that a fairly simple mathematical sum will give the distance between your finger and your face.

This is what Henderson did, from the Cape of Good Hope Observatory, in the 1830s. He needed a much longer baseline than the distance between his eyes, so he used the diameter of the Earth's orbit round the Sun, known to be 186,000,000 miles. He believed that Alpha Centauri must be fairly close by cosmical standards, and so he observed first in January and then in July, by which time the Earth had moved round to the opposite side of its orbit. He could now measure the parallax simply by finding the apparent change in the position of Alpha Centauri relative to the more distant background stars, and he finally gave the distance of Alpha Centauri as just over four light-years.

Actually, Henderson missed his main chance of immortality. He had become His Majesty's Astronomer at the Cape, but he hated the place and even referred to it as 'a dismal swamp'. When he resigned and returned home, he did not hurry to work out his results, and before he published them he had been forestalled by the German astronomer Friedrich Bessel, who used exactly the same method to work out the distance of a faint northern star, 61 Cygni in the constellation of the Swan. No matter; Henderson came to the right answer, and now we can look ahead of us to see what the Alpha Centauri system is like.

The faint member of the trio, Proxima, is the nearest of all the stars beyond the Sun. Quite obviously it is one of Nature's stellar glow-worms. Instead of a bright orange or yellow surface it is red, which means that it is considerably cooler than the Sun. It is also much smaller, with a diameter of only about 40,000 miles. Proxima is smaller than the largest planets in the Solar System, Jupiter

Impression of Tau Ceti, a star not too unlike the Sun. Eleven light years away from us, it is a possible centre of a planetary system. We see here a view from the satellite of a planet in the Tau Ceti system.

Bennett's Comet – seen in 1970; one of the wanderers who will not return to the Sun for many centuries. In our journey we may meet it on its way back to the Oort Cloud, but it will have lost its splendour – and its tail.

and Saturn, but it is much more massive, and it is a true star even though it shines with only about 1/13,000 the power of the Sun. Indeed, if Proxima suddenly replaced the Sun at the centre of our Solar System it would send the Earth only as much light as 45 full moons, and the temperature would drop so quickly that all living things would be frozen to death.

Closing in, something else of great interest catches our attention. Proxima is not shining steadily. There are occasional bursts of abnormal brilliance, as though the dim red star were suffering violent, short-lived outbreaks. From our vantage point we can see that this is indeed true, but the outbreaks or flares are limited to small areas of the star's surface, and they do not last for more than a few minutes. Our Sun shows flares of this kind, generally associated with large spot-groups, but the Sun is so luminous that a flare makes no detectable difference to its total output, whereas with the feeble Proxima the flare is very evident. Many red dwarfs show the same flare phenomenon, though we cannot yet pretend that we can explain it in detail.

Proxima is a star that has never shone as brightly as the Sun, because its mass is too low, and it will cool down very slowly until its light and heat leave it. Neither can it be brilliantly lit by the Alpha Centauri twins, because they are at least one-sixth of a light-year away – that is to say, 10,000 times the distance between the Sun and the Earth – and though they dominate the sky, they send very little heat out as far as Proxima. Not that this is important, because Proxima has no planetary system, and neither can we expect to find planets orbiting the main Alpha Centauri pair, because the distance between them is too slight. Moving on, we find that one component is larger than the other, but it is the smaller star which is the more luminous; it somewhat outshines the Sun, and is of the same yellow hue, while the larger secondary component is orange, indicating a lower surface temperature.

Alpha Centauri is an excellent example of a binary system, as any small Earth-based telescope will show (not from Britain, alas, because Alpha Centauri never rises above the horizon). The two components are moving round their common centre of gravity, rather in the manner of two dumb-bells being twisted by their joining arm, taking almost 80 years to complete a full orbit. This orbit is not circular. At their closest, the two stars are little over 1,000 million miles apart; at their furthest, over 3,000 million miles. Compared with the scale of the Solar System, this means that the minimum separation is a little more than that between the Sun and Saturn, while the maximum is greater than the distance between the Sun and Neptune. Any planet would have a wildly erratic orbit, and the temperature range would be extreme.

Looking back towards the Solar System we notice something else. From Alpha Centauri the constellations look very much the same as they do from Earth, and it is easy to pick out the W of stars which marks Cassiopeia and is familiar to anyone who lives in the British Isles, because it is so far north that it never sets, and is always visible whenever the sky is sufficiently dark and clear. But now the W has been joined by a bright star of the first magnitude, rather less brilliant than Alpha Centauri itself appears

to us. The newcomer is the Sun, seen in its true guise as an ordinary yellowish star. From four light-years we would need a very powerful telescope to see any of the Sun's planets, even Jupiter; the Earth itself would be hopelessly lost.

If there are any planets in the Alpha Centauri group, we still cannot hope to find life. It is time to journey on, and another two light-years will bring us to a red star known officially as Munich 15040, but more generally as Barnard's Star, since it was first studied many years ago by the keen-eyed American astronomer Edward Emerson Barnard. Like Proxima, Barnard's Star is a red dwarf. It is less of a glow-worm, and has 1/2,500 the luminosity of the Sun; its surface temperature is 3,200 degrees, a little more than half that of the Sun, and its diameter is less than twice that of Jupiter.

The active Sun. A huge cloud of hydrogen gas rises from the Sun. Prominences are often associated with flares – but as we approach Proxima Centauri we can see that the flares there, unlike those of the Sun, visibly change the total output of light.

If Barnard's Star is so like Proxima, then why should we stop to examine it more closely? The reason is that there may well be an associated planetary system, and there is even a certain amount of evidence that Barnard's Star is not a solitary wanderer, though it is not a member of a binary system.

The star is remarkable because of its exceptionally great individual or proper motion. The stars are not genuinely fixed; they are moving in all sorts of directions at all sorts of speeds, but they are so far away from Earth that their proper motions are too slight to become obvious except over immensely long periods. King Canute, Julius Cæsar and even Homer would have seen the constellations almost exactly as we do. However, Barnard's Star shifts relatively quickly against its background. In only 180 years it moves by a distance equal to the apparent diameter of the full moon, and this is easy to measure, so that the star is often nicknamed 'the Runaway'.

At the Sproule Observatory in America Peter van de Kamp and his colleagues have been paying special attention to Barnard's Star. It is not moving in a regular straight line, but wobbles its way along. Each wobble is very slow and very slight, but it does indicate that the star is being pulled upon by an invisible companion. The companion cannot be a star; it is not massive enough for that, and so it is presumably a planet. Van de Kamp believes that there may be at least two planets, one almost as massive as Jupiter and the other comparable with Saturn. And if there are two members in the system there may well be others, too small and lightweight to produce any observable effects.

Let us take a careful look. The planets – if they exist – are bound to be gloomy places,

Impression of the Alpha Centauri System. We approach Proxima, the Red Dwarf component; brilliant flares common to this type of star rise from its surface. The double star Alpha, though ⅕ of a light-year away, dominates the sky.

since they receive so little light and heat from their parent star; they are bitterly cold, and their daylight will be nothing more than an eerie red glow. Life there could not evolve in the same way as has happened on Earth, but it would be rash to jump to the conclusion that there can be no life at all. Our present knowledge is far from complete, and we know that living things can flourish in the most unlikely places.

There may be a chance of learning more during the next few years. In 1985 the Space Telescope is due to be launched. It will have a 94-inch mirror to collect its light, and since it will operate from the top of the Earth's dirty, unsteady atmosphere it will have unrivalled power. It could even show the planets of Barnard's Star. Remember that 'the Runaway' is only six light-years away from us, and only the Alpha Centauri system is closer than that.

Next on our list comes a star, or rather a star-pair, of very different type: Sirius. From Earth it shines as the brightest star in the whole sky, and is magnificent during winter evenings in Britain; look for it rather low in the south, in almost a direct line with the three stars of Orion's Belt. It twinkles violently, and flashes different colours, sometimes red, sometimes green, so that it can hardly be mistaken. But from our vantage point in space, there is no twinkling. The twinkling or scintillation of a star is due solely to the Earth's air, and once above the air the twinkling stops, so that Sirius shines steadily with a brilliant, pure white radiance.

For the first time we have come to a star which is much more powerful than the Sun; it would take 26 Suns to match one Sirius, and the size and mass are also greater. Sirius has a diameter of over 15,000,000 miles and is more than twice as massive as the Sun, so that it is using up its reserves of hydrogen 'fuel' at a much faster rate. It is two and a half light-years further away than Barnard's Star

– that is to say, rather more than eight light-years away from the Earth – but it still ranks as a very near neighbour.

As the range decreases we see another star, only 1/10,000 as bright as Sirius itself, but clearly associated with it. This is Sirius B. Since Sirius is often nicknamed the Dog-Star, the companion is referred to as the Pup, but it is a very heavy pup. Tiny though it is, with a diameter of only about 26,000 miles (less than that of a planet such as Uranus or Neptune) it has almost the same mass as the Sun. The distance between the Dog and the Pup is about 2,000 million miles, and they revolve round their common centre of gravity in a period of 50 years.

If the two stars were equal in mass, the 'balancing point' or common centre of gravity would be exactly half-way between them. Since Sirius A is two and a half times the heavier, the balancing point is shifted markedly in its direction, but this means that irregularities in the star's proper motion will show up, and it was these which led Friedrich Bessel to suggest that a faint companion star should exist. In 1862 Alvan G. Clark, one of the leading instrument-makers of his time, was testing a new telescope when he caught sight of the Pup, almost exactly where Bessel had expected it to be.

The Pup proved to be a problem. Its mass could be found because of the way in which it and the Dog-Star moved. If it were equal in mass to the Sun, and yet very faint, it would presumably be large and red; at least, so astronomers reckoned. It came as a major surprise when it was examined in detail in 1915 by another American, W. S. Adams, who used a spectroscope fitted to a very large telescope. The Pup was not cool and red, but hot and white; its surface temperature

Impression of a planet of Barnard's Star. Because of its motion through the sky, astronauts suspect that Barnard's Star may well have large planets in orbit around it. The scene is eerie, dimly lit by the small Red Dwarf Sun.

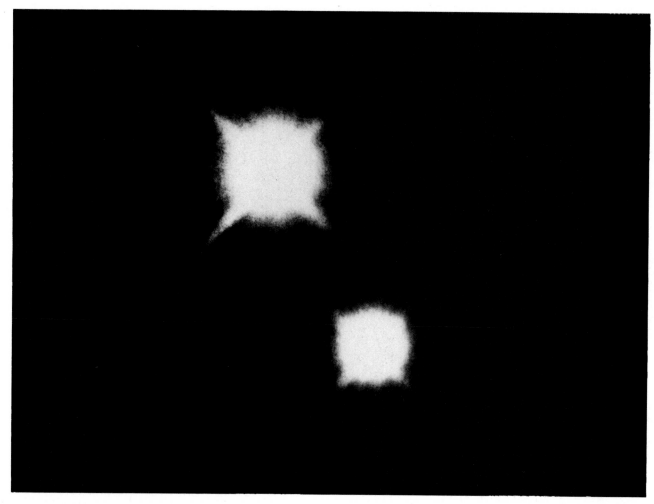

The Pointers. One of our first stops is at Alpha Centauri, the nearest star-system to the Sun. We can also see Beta Centauri, the second pointer to the Southern Cross, but we must journey for hundreds of light-years before reaching it.

proved to be 8,500 degrees centigrade, much higher than that of the Sun.

The situation became even more curious. If the Pup were hot, white and faint, it would have to be very small, in which case its density would have to be improbably high – certainly, at least 60,000 times that of water. An Irish astronomer, J. Ellard Gore, commented that since such a density was obviously absurd, there must be a major mistake in the observations.

Yet there was no error. The Pup really is of this density, so that a cubic inch of its material would weigh two and a quarter tons. The theorists set to work, and within a few years they had given a correct explanation. The Pup is a white dwarf, which has come near the end of its evolutionary sequence and has no hydrogen 'fuel' left.

Inside our Sun, energy is being produced by nuclear transformations. During our journey to the solar core we saw how the nuclei of hydrogen atoms are banding together to make nuclei of helium, energy being emitted in the process. Æons ago the Pup shone in the same way, but when it had used up its supply of hydrogen it was forced to rearrange itself. First it swelled out to become a large, red star; then it threw off its outer layers, and finally its remaining

Sirius. A brilliant white star; from Earth it appears the brightest star in the sky. From our vantage point in space we see that Sirius shines with a steady, pure white light. Its companion, the Pup, is 10,000 times less luminous than Sirius itself.

material collapsed, so that the atoms were crushed and broken. This meant that they could be crammed together with very little waste space, producing material which is incredibly dense. This is the state of the Pup today. It is bankrupt, and can never regain its former glory; it will merely go on shining very feebly, by gravitational energy, before it loses the last of its light and becomes a cold, dead globe or black dwarf. Of course, the process will not be rapid; it will take thousands of millions of years. By then the Dog, too, will have gone through the same series of changes, and instead of the present system, with one brilliant star and one dim one, we will be left with a white dwarf Dog and a dead Pup.

Planets are unlikely to exist. A binary pair is not a suitable candidate for a planetary system. We must look for a single star which is not too unlike the Sun, and for this we must move further away to about eleven light-years, where there are two possible candidates, Tau Ceti and Epsilon Eridani. Both are visible with the naked eye from Earth, though both are somewhat cooler and redder than the Sun. We have a straight choice, so let us select Tau Ceti, which has less than half the luminosity of the Sun and a diameter of 400,000 miles.

Not even the Space Telescope will show Tau Ceti as anything more than a speck of light, but in our imaginary journey we can come close to it, and we see a star which is not too unfamiliar, though the lower surface temperature means that the colour is orange rather than yellow. Looking around, we may reasonably expect at least one planet, though whether it (or any others in the system) will be at all like the Earth is quite another matter.

We can learn a great deal from thinking about the way in which our own Solar System was created. Older theories involved an encounter between the Sun and a passing star. Sir James Jeans, well remembered today both for his theoretical work and for his popular books and broadcasts, worked out and refined a picture according to which a wandering star approached the Sun and pulled a huge, cigar-shaped tongue of matter away from the solar surface; as the intruder withdrew, the 'tongue' was left whirling round the Sun, eventually breaking up into blobs which became the planets. It was significant that the largest planets, Jupiter and Saturn, lay in the middle of the Solar System, where the thickest part of the cigar would have been.

It all looked very neat and tidy, but before long serious objections were found. Mathematical analyses showed that even if such a cigar could be forced out of the Sun, it would not break up in a manner which would lead to a few definite planets, and attempted

The Space Telescope will be the first of many such instruments, flying freely in space and able to examine the remote bodies of the universe beyond the Earth's dirty, unsteady atmosphere.

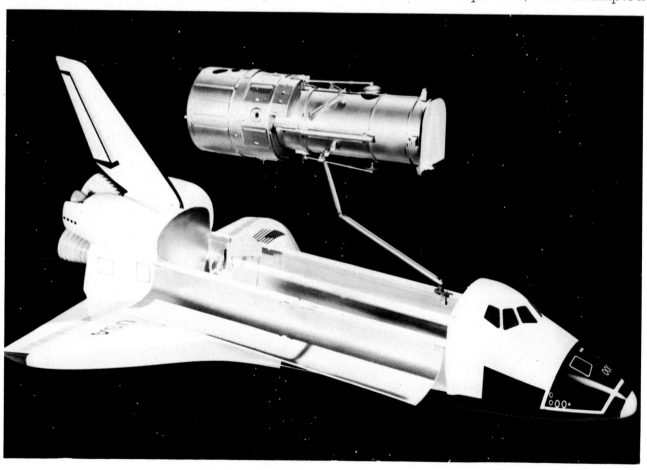

Looking inwards, we know that the centre of the Galaxy lies beyond the star clouds in the Milky Way, in Sagittarius, but we cannot see through the dust, and we have to rely upon infra-red and radio waves to bring us information about the true centre.

modifications were no more successful. Other theories involved a binary companion of the Sun which exploded and then drew away, planet-forming material being scattered in the process. These, too, were rejected, and astronomers went back to a theory which was not too unlike the 'Nebular Hypothesis' described by the French astronomer Laplace as long ago as 1796.

Laplace began by assuming that the Sun condensed from a cloud of gas and dust in space. We know of many clouds of this kind, and call them nebulæ, so that there was nothing improbable in such an idea. The solar nebula would shrink under the influence of gravity, and Laplace believed that as it did so, it would throw off rings of material, each of which produced a planet – in which case the outermost planets such as Neptune and Uranus would be older than those which are closer in.

Current theories follow the same general trend, though without involving Laplace-type rings. It is now thought much more probable that as the solar nebula contracted, the planets grew by the process of accretion. Far away from the Sun, the material was very cold, and planets such as

The 'Pulsar' radio telescope at Cambridge. Outwardly it is anything but impressive, but it led in 1967 to one of the most important discoveries in modern astronomy; the existence of ticking sources or pulsars.

Jupiter and Saturn built up, pulling in light gases such as hydrogen and helium as soon as they became massive enough. This is why the giants now contain so much hydrogen. Nearer the centre of the cloud the lightest gases were driven out, and the only solids were those which now make up the inner planets, including the Earth. The remnant of the solar nebula is nothing more nor less than the present Sun.

Of course, this description is hopelessly over simplified, but as we draw in towards Tau Ceti it will serve us quite well. On the Jeans passing-star theory, planetary systems would be very rare. Space is thinly populated, and the chances of a stellar collision or even a 'close encounter' are very remote. But if systems of planets are produced in the way now generally believed, planets are likely to be common. What can happen to the Sun can happen to other stars, too; there is no reason for our Sun to be singled out.

However, we must look for restrictions. A very luminous star will run through its life-cycle quickly, so that even if planets are formed there will be insufficient time for life to evolve upon them before conditions become intolerable. Binaries are also unsuitable (unless the components are very widely separated), and so are the variable stars, which, as their name implies, change in brightness and output over short periods. We must concentrate upon single stars of roughly the same type as the Sun – and this brings us back to Tau Ceti.

Unfortunately, even though eleven light-years is trifling by the standards of the Galaxy, it is too far for us to have any hope of seeing a planet associated with Tau Ceti. The only possible way to establish the existence of life there is by radio, remembering that radio waves move at the same speed as light.

Radio astronomy began in the 1930s, with some work by an American engineer named Karl Jansky. He was investigating 'static' on behalf of the Bell Telephone Company and had built a large aerial, made partly from bits of a dismantled Ford car. He found, to his surprise, that he was picking up radio waves from the Milky Way. For some reason or other he never followed up this epoch-making discovery, but after World War II radio astronomy became a vitally important branch of science, and huge radio telescopes were built, the most famous of which is the 250-foot 'dish' at Jodrell Bank in Cheshire, masterminded by Professor Sir Bernard Lovell.

Two points must be borne in mind: first, heated bodies in the sky give out radiations at all wavelengths – not only visible light – and so the radio signals which we receive are natural, not artificial; second, a radio telescope collects and concentrates radio waves just as an ordinary telescope collects and concentrates light, but no visible picture is produced, and one cannot look through a radio telescope. The information is generally given in the form of a tracing upon graph-paper fixed to a rotating drum. Not all radio telescopes are on the same pattern as Jodrell Bank; others look more like collections of barbers' poles, though all are doing essentially the same thing.

Many radio sources were detected, but in 1961 a group of astronomers at Green Bank, West Virginia, began a novel experiment. They used their 85-foot dish-type radio telescope to listen out at selected wavelengths in the hope of picking up signals rhythmical enough to rule out any natural origin. The two selected target stars were, predictably, Tau Ceti and Epsilon Eridani. Nothing unusual was found, and though this first attempt – known officially as Project Ozma, more commonly as Project Little Green Men – was soon abandoned, other similar experiments have been made since.

Even if we could transmit a rhythmical signal to Tau Ceti, and if an obliging radio astronomer living on a planet in that system heard it and replied at once, we would still be restricted. Our transmission would take eleven years to reach Tau Ceti, and the reply would take a further eleven years, so that if the transmission were sent out in 1983 there would be no chance of a reply before 2005, making quick-fire repartee rather difficult. Yet it would establish the fundamental fact that there really is intelligent life elsewhere in the Galaxy, and this would have profound effects upon our thinking, both scientific and philosophical – even religious. And since mathematics is universal (we did not invent mathematics; we merely discovered it), there would be every hope of establishing intelligible contact eventually.

Tau Ceti and Epsilon Eridani have remained silent. Perhaps we are concentrating upon the wrong stars; after all, both are much less luminous than the Sun. Therefore, we will return to our light-beam and continue our journey for another eight years until we find ourselves approaching the star Delta Pavonis, which is almost a carbon copy of the Sun. It is only slightly smaller, less massive and less luminous; it is yellow, with a surface temperature the same as that of the

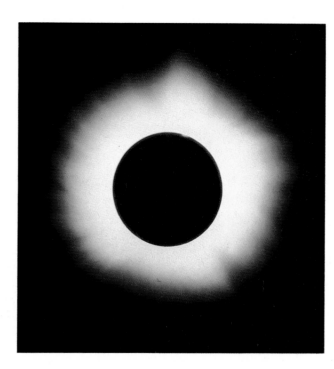

Total eclipse. From Earth, our Moon just covers the Sun, producing a total eclipse and showing the beautiful corona. Does Delta Pavonis have such a planet with such a moon? If so, will inhabitants of that world see total eclipses of *their* sun?

Sun, and its distance from the Earth is nineteen light-years.

To us, Delta Pavonis looks unremarkable enough. It is never visible from the British Isles, because it never rises; the constellation of Pavo, the Peacock, is too far south in the sky. But it is so sunlike that we are entitled to be optimistic. It may so easily have planets of its own.

On the other hand, we must again be wary of assuming that a planetary system must necessarily harbour intelligent life. In our Solar System only the Earth is suitable. The other planets are either too hot or too cold, or lack breathable atmospheres. If Delta Pavonis has an inhabited planet, then the planet must be comparable in size with the Earth, and must move round its sun at a distance ranging from around 80,000,000 to 100,000,000 miles. There is absolutely no guarantee that conditions such as this would be fulfilled.

However, let us assume that there is such a planet. What will the 'Pavonians' be like?

Circling this sun-like star, we must think a little more deeply about what 'life' entails. At the moment we are still very much in the dark. We cannot create life (though, unfortunately, we can destroy it with alarming ease). Yet we know a great deal about the material which makes up living things, and we know that everything depends upon the unique properties of one type of atom, that of carbon. Only carbon atoms can join up with others to produce the complicated atom groups or molecules which are needed. This is fact, not speculation. We have shown that the materials making up the universe are the same everywhere; the stars, the star systems and the remote galaxies contain atoms of carbon, hydrogen, oxygen and all the rest. We have a complete list of elements. Ninety-two types occur naturally, hydrogen being the lightest and uranium the heaviest. Still heavier elements can be produced artificially, but all these are unstable, and as life-producers we can rule them out.

Therefore all life, whether here, on Mars, on a planet of Delta Pavonis or in a distant galaxy, must be carbon based. It may not look like ourselves; I am quite prepared to believe in an intelligent astronomer on some faraway world who is equipped with two heads and six arms, but he will still be of a type which we can understand, and therefore will not be what science-fiction devotees call a BEM or Bug-Eyed Monster, although he (or she, or it) may look like one.

Of course, there may be a fundamental flaw in this reasoning. It is impossible to prove a negative. I cannot prove that from my home on Selsey Bill the Sun will rise in the east tomorrow morning, but I think it will, because all the evidence points that way. Faced with a set of facts which is admittedly incomplete, all that can be done is to arrange them and then interpret them as logically as possible. And the evidence available to us today indicates that all life must be carbon based. Unless or until any contrary evidence comes to hand, we must restrict ourselves to 'life as we know it', otherwise, speculation becomes not only endless, but also pointless.

By now we are nearing Delta Pavonis; we see a yellow star with spot-groups here and there, looking remarkably like our Sun. We look for a planet in the region round the star where the temperature is neither intolerably high nor intolerably cold. If we are lucky, there may be such a planet, similar to the Earth in size and mass, possibly with one or more satellites. It is here that we must search for life.

Life on Earth began in the oceans, some time after the Earth itself came into existence about 4,700 million years ago. Most scientists believe that it originated there, but there are some dissentients, notably Sir Fred Hoyle, who considers that life-bearing material was dumped upon Earth by a comet, and Francis Crick, Nobel Prize-winner and one of the world's most famous scientists, who goes so far as to suggest that the first living things were deliberately sent to our world by an intelligent civilization far across the Galaxy. But whatever may be the truth of this, life on an Earth-like planet of Delta Pavonis might be expected to evolve along familiar lines. There is nothing absurd in suggesting that the Pavonians might resemble ourselves, with man-like virtues and weaknesses. Perhaps they, too, have built their radio telescopes, and have started to listen out for rhythmical signals from the yellow star in their sky which we call the Sun.

It would be a happy coincidence if life developed on two planets, separated by only nineteen light-years, at the same time. Even if Delta Pavonis does have a suitable planet (and let us admit, once again, that of this there is not the slightest proof) its

The great 250-foot radio telescope at Jodrell Bank. From here astronomers carry out investigations of bodies so far away that their light takes many millions of years to reach us – worlds which we plan to visit on our light-beam.

inhabitants may still be in a Stone Age. Alternatively, they may have become much more advanced than we are, or else wiped themselves out in a nuclear holocaust. We cannot tell, and at present we have no hope of finding out. All we can claim is that all of these speculations are possible.

We are moving out from the Sun's immediate neighbourhood into the further reaches of the Galaxy, and it is time to equip ourselves with extra instruments. As we travel on, covering 186,000 miles every second, Delta Pavonis, Tau Ceti, Epsilon Eridani, Sirius, Alpha Centauri and the Sun fade into the distance – and yet we have barely started the main part of our journey.

6 The Suns of Space

Delta Pavonis is far behind us. On our light-beam, we have left all our familiar surroundings, and at a distance of 26 light-years we come up to one of the loveliest and most famous stars in the sky: Vega. From the British Isles it is almost overhead during summer evenings, and it is striking both because of its brightness and because of its steely blue colour. Blue means hot; evidently Vega has a high-temperature surface, and indeed this is true. The temperature is over 9,000 degrees centigrade, and Vega is more than 50 times as powerful as the Sun. For once we are dealing with a single star. From Earth, Vega seems to have a faint companion, but this is merely a line-of-sight effect. Vega is not a binary; the faint companion is in the background, and simply happens to lie in much the same direction as seen from Earth.

We have allowed ourselves to travel across the Galaxy at the velocity of light — something which, so far as we can tell, no material body can ever do. Having taken this liberty, let us take another, and assume that we are carrying a telescope of immense power, capable of looking back at the Earth and seeing its surface features in detail. Vega is 26 light-years away. Therefore, when we turn our conjured-up telescope Earthwards, we see the scene not as it is today, in 1983, but as it used to be 26 years ago. The date is 1957, and Sputnik 1, the

The Vela supernova remnant. We see a portion of a roughly circular shell of filaments in the southern constellation of Vela. These outline the position of the still spreading blast wave from the detonation of a supernova some 10,000 years ago.

world's first artificial satellite, is just about to be launched.

Frankly, the idea of a telescope capable of this sort of viewing is even more far-fetched than moving along on a light-beam, but it will be convenient to us for our present purpose, so let us christen it the Super Telescope. It will help us a great deal, and we will use it to the full, though remembering that nothing of the sort has ever been built — and never will be!

Much has been heard about 'time-travel'. All the evidence indicates that travelling backwards in time, in the manner of Dr Who, is one of the few things which we may dismiss as genuinely impossible, but we can certainly look backwards in time, which is the next best thing. In 1983, we on Earth see Vega as it used to be in 1957. Similarly, any inhabitant of a planet moving round Vega could use his Super Telescope to see the Earth in its 1957 guise; and if he looked at exactly the right moment, he would see the lift-off of Sputnik 1. He could even hear the famous 'bleep! bleep!' signals sent out by the football-sized satellite as it sped round the world.

Even Vega is close on the cosmical scale, and as we move further and further out our Super Telescope will show the Earth as it used to be longer and longer ago. It is also clear that our Earth-based view of the universe is bound to be out of date, and we see the stars as they were at different stages in their careers. We can even see stars which no longer exist. But this can come later; meanwhile, what about Vega itself?

Sputnik 1. The first artificial satellite, which opened the Space
Age on 4 October 1957 – so that our Super Telescope at the
distance of Vega will now be able to re-observe it!

It is larger than the Sun, with a diameter of
about 2,700,000 miles, and its mass is three
times that of the Sun. Because it is luminous
and energetic, it is squandering its fuel
supply at several times the solar rate, and so
it will not last for so long, but at the moment
it is in a stable condition, and nothing will
happen to it in the foreseeable future. It is
what is termed a Main Sequence star.

Over 80 years ago, two astronomers –
Ejnar Hertzsprung of Denmark and Henry
Norris Russell of the United States – made
independent studies of the colours, temperatures and luminosities of the stars.
Their main equipment was the spectroscope, which is important in astronomical

science. Broadly speaking, a telescope
collects light, while a spectroscope splits it
up and tells us which chemical substances
are present in the light-source (which is how
we know that the stars are made up of
familiar elements, notably hydrogen and
helium). The stars had already been divided
up into various spectral types, each denoted
by a letter of the alphabet. It would have
been logical to use the A, B, C . . . sequence,
but, as so often happens, the original system
was found to be faulty, and eventually the
sequence became alphabetically chaotic. It
has become O, B, A, F, G, K, M in order of
decreasing surface temperature, O stars
being the hottest and M the coolest. (Note
the famous mnemonic O Be A Fine Girl Kiss
Me.) Stars of types O and B are bluish or
white; A, white; F, yellowish; G, yellow; K,

orange, and M orange-red or red. (A few extra classes have also been included, but they contain small numbers of stars, and for the moment we need not bother about them.) Vega is of Type A while Delta Pavonis and the Sun belong to Type G, Tau Ceti and Epsilon Eridani to Type K, and Proxima Centauri and Barnard's Star to Type M.

Hertzsprung and Russell then proceeded to draw up diagrams in which they plotted the stars according to their surface temperatures (or their spectra, which comes to the same thing) and their luminosities. They

found that on such a diagram, known today as an H-R Diagram, the distribution was not random. Most of the stars fell near a line running from the upper left to the lower right, and it is this which makes up the Main Sequence.

Even more remarkably, the red and, to a lesser extent, the orange stars fell into two distinct groups. They were either very dim or else very brilliant; there seemed to be no intermediate stars. It was natural to term these the Giant and Dwarf branches. Red dwarfs such as Proxima fall into the Main Sequence; red giants such as Betelgeux in Orion, which we will visit shortly, do not. (White dwarfs were not known when the first H-R Diagrams were drawn up, but they

The H-R Diagram. Perhaps the most important diagram in modern astrophysics, linking a star's luminosity with its spectral type. Most stars lie on the Main Sequence. Note also the White Dwarfs, which are nearing the end of their careers.

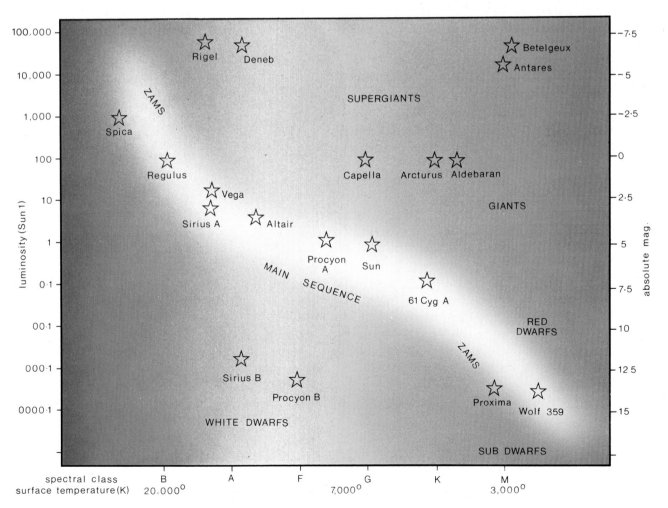

do not fall into the general pattern. They have long since left the Main Sequence, and are quietly dying.)

Not surprisingly, it was thought that an H-R Diagram presented a picture of how a typical star evolved. It would condense out of dust and gas as a cool red giant; as it shrank, under the influence of gravity, it would heat up and join the top of the Main Sequence, after which it would gradually slide down the Sequence towards the bottom right before ending its career as a faint red dwarf. It looked beautifully simple, but it turned out to be wrong. Things are much more complicated. However, both Vega and the Sun will stay on the Main Sequence for long periods. The Sun will remain there for another 5,000 million years at least, Vega for considerably less.

Vega is a glorious star, but probably it is too hot and energetic to have a planetary system, so we will bid it adieu and move out for another ten light-years. Now we are 36 light-years from the Earth, and are closing in on another brilliant star, Arcturus in the constellation of the Herdsman, though by now the familiar constellation patterns have become distorted because we have moved sufficiently far from our starting-point on Earth. Arcturus, too, is beautiful, but in a different way from Vega. Instead of being blue, it is light orange, so that it is cooler; but it is also larger, with a diameter of some 20,000,000 miles, so that its total luminosity is over three times that of Vega, equalling 115 Suns. As befits an orange star, its spectrum is of Type K.

As we draw near, a tremendous flood of radiation engulfs us; Arcturus is a powerful star. Yet on Earth, at a range of 36 light-years, we can detect almost no heat from it. Calculations show that the heat from

Impression of Betelgeux. Sunrise on a planet in orbit around this red giant star is spectacular. As Betelgeux explodes into a supernova outburst, there will first be a blast of radiation and then numbing cold as the star becomes very feeble.

Arcturus is about the same as that of an ordinary candle at a distance of five miles.

Arcturus is one of the few bright stars with a proper motion well above the average (Sirius is another). At present it is approaching the Earth and the Sun, but there is no danger of a future collision, because Arcturus will not continue its approach indefinitely. It will begin to move away again, and in 500,000 years' time it will be so far away from the Earth that it will have become invisible with the naked eye. However, from our light-beam spacecraft it is imposing by any standards.

Using our Super Telescope to look back at the Earth, we see things as they used to be in 1947, when the war was not long over and many European cities were in ruins. We might see Winston Churchill walking to or from Westminster; cricket-lovers might enjoy the spectacle of seeing Compton and Edrich tear the South African bowling to shreds in that long-forgotten, gloriously hot

The surface of Betelgeux. As we draw in towards Betelgeux, we can see patches on its orange-red surface; they have also been detected from Earth by a new technique known as speckle interferometry.

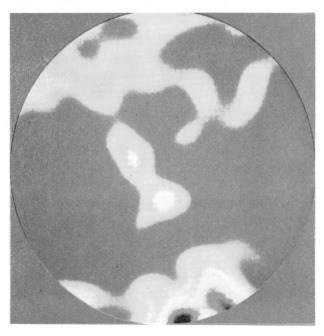

and peaceful English summer. In the United States, Harry S. Truman was President, having won his unexpected victory; Franklin D. Roosevelt had died only two years before. But Arcturus is well past its prime, and has left the Main Sequence. If it has planets, they are not likely to be inhabited. Arcturus is much more luminous now than it used to be in its Main Sequence days, and any planets moving round it would have had their climates drastically altered.

While we are still 36 light-years from Earth, it will be worth taking a look at the binary system of Arich, or Gamma Virginis, which is totally unlike the Sirius pair. Viewed from Earth, Arich seems like an ordinary, moderately bright naked-eye star, but a telescope shows it to be made up of two stars, almost exactly equal in every respect. Both are yellowish, with surfaces slightly hotter than that of the Sun, and both have about three and a half times the Sun's luminosity. They take 180 years to complete one journey round their 'balancing point' or common centre of gravity, but they are not always the same distance apart; their separation ranges between 6,500 million miles and only 280,000,000 miles.

From Earth, they are not so striking now as they used to be a few decades ago, because we are seeing them from a less favourable angle; one star is moving more or less behind the other, so to speak, and by the year 2010 a large telescope will be needed to show them separately. On our light-beam we are not so restricted. 'From above', for instance, both stars will be seen at their best; but what about the conditions on a planet attending one of them?

Suppose that one component star (A) has a planet moving round it at a distance of 36,000,000 miles, which is the real distance between the Sun and its closest planet, Mercury. When Arich A and its companion (B) are at their closest, A's planet will be scorched not only by its own sun but also by

Mizar, the celebrated multiple sun. When this picture was taken, in 1975, a comet – Kobayashi-Berger-Milon – was in the same region of the sky; it is seen above the Mizar pair.

B, which will be less than 280,000,000 miles away. There can be no darkness; as one sun sets the other rises, and the combined heat of the two will be colossal. Certainly any atmosphere of the hapless planet will long since have been stripped away, and it can be no more than a sterile ball. Even in the conditions 90 years later, when B has receded to its maximum distance, the temperature of the planet will still be high, though B will appear as a blindingly brilliant speck instead of a huge disk. Conditions might then be less intolerable, but not for long; as B drew inwards again the heat would climb steadily back towards the fiercest period of the planet's strange summer. Whether any such planet really exists must be regarded as highly doubtful.

If it does, it is hardly likely to welcome visitors. So let us move outwards again, until we have reached a distance of 42 light-years from the Earth. Our next target is Capella.

Capella is overhead from Britain during winter evenings, when it takes the place occupied during summer evenings by Vega, but the two are not alike. Vega's steely blue contrasts with the pale yellow of Capella. It might be thought that Capella would be similar to the Sun – but whereas the Sun is a Main Sequence star, Capella is a giant. More precisely it is two giants, since Capella proves to be a close binary. From Earth the components merge into one; but as we approach on our light-beam we find that there are two yellow giants, so close together that their surfaces are less than 60,000,000 miles apart, moving round their balancing point in only 104 days. Both are much larger than the Sun, with diameters of 11,000,000 miles and 6,000,000 miles respectively.

Close to Capella there is ceaseless turmoil. The two components are so close together that they raise vast tides in each other, and material is scattered around; it would be a fascinating spectacle, but one to be viewed from a respectful distance. The likelihood of a planetary system seems small; a planet of the Capella twins would have to be a long way out, and would hardly be of the type suited to our kind of life.

Looking back at the Solar System from the distance of Capella, we find that the Sun has become too faint to be seen with the naked eye. With our Super Telescope, the Earth would be seen as it was in the grim days of 1940–41, when Hitler's aircraft swarmed to the attack and the Battle of Britain was in full swing. We could see again the deadly dogfights, the bombing of London and other cities, and the whole senseless orgy of destruction. Moreover, with a radio receiver of similarly limitless sensitivity, we could hear Winston Churchill proclaiming that we would never surrender. If we wished, we could also hear Hitler and Göring declare that the war was about to end in a great victory for the Third Reich.

Note that from Capella, the Solar System would be 'radio noisy'. There would be not only the natural long-wave emissions from the Sun (and Jupiter) but also the artificial broadcasts from Earth. Regular broadcasting began in around 1920; before that, our transmissions were infrequent and weak. This means that the Earth is now radio noisy out to a distance of just over 60 light-years. The broadcasts from 2LO will have passed Capella, and will have reached out to our next target, Mizar in the Great Bear, whose distance is 59 light-years. As yet, our

Impression of Castor. A planet of a star which is made up of six components; two bright pairs and one faint pair – Castor, the senior though fainter member of the Twins in Gemini, near Orion. (The other twin is the orange single star Pollux.)

broadcasts have not had enough time to penetrate further into space. From, say, the star Regulus, at 85 light-years, the Earth will still be radio quiet. If there are any listeners there, they will start to hear us about the year 2000. Before that, they can have no inkling of any artificial transmissions from the Solar System.

Mizar is more than a simple binary; it is a kind of family party of stars. From Earth it is seen as the second star in the tail of the Great Bear (or the handle of the Plough); it is fairly

The Hyades are drowned by the bright orange light of Aldebaran: but Aldebaran is not a true member of the cluster. It simply happens to lie in the same direction as seen from Earth, and is only half as distant.

bright without being outstanding, and close beside it is a fainter star, Alcor, easily visible without optical aid on a clear night. In the year 1651 an Italian astronomer named Riccioli found that Mizar itself is double, and since then each component has proved to be again double, so that we have four main components in the system, plus Alcor, which is itself a close binary.

Swooping in towards the Mizar group, we will see it in all its complexity. Of the two main components, one (A) is made up of two stars, each 35 times as luminous as the Sun, moving round each other in only twenty and a half days at a separation of 18,000,000 miles, less than the distance between Mercury and our Sun. At a distance of 33,000 million miles we come to the second bright binary (B), where the two components are rather less powerful and further apart, with a revolution period of 182 days. No planets can be expected; if any existed, their orbits would indeed be variable and erratic. Finally there is Alcor, a quarter of a light-year from the main pair. Presumably it shares in the general revolution of the system, but it must take millions of years to complete one circuit. Feeble though it looks from Earth, it is still fifteen times as luminous as the Sun.

It used to be thought that a binary pair began as a single star, which rotated so rapidly on its axis that it became unstable and finally broke in two, but nowadays it seems more probable that stars really are born in clusters, so that the components of a binary have always been separate. If so, the four Mizars and the two Alcors have a common origin, and are of about the same age, probably younger than the Sun, because they are hotter and more energetic. All six stars are white, and with Castor, in Gemini, we have another sixfold system — two bright white pairs and one faint red pair.

With Mizar, we have reached the limit of Earth-transmitted radio broadcasts, and at

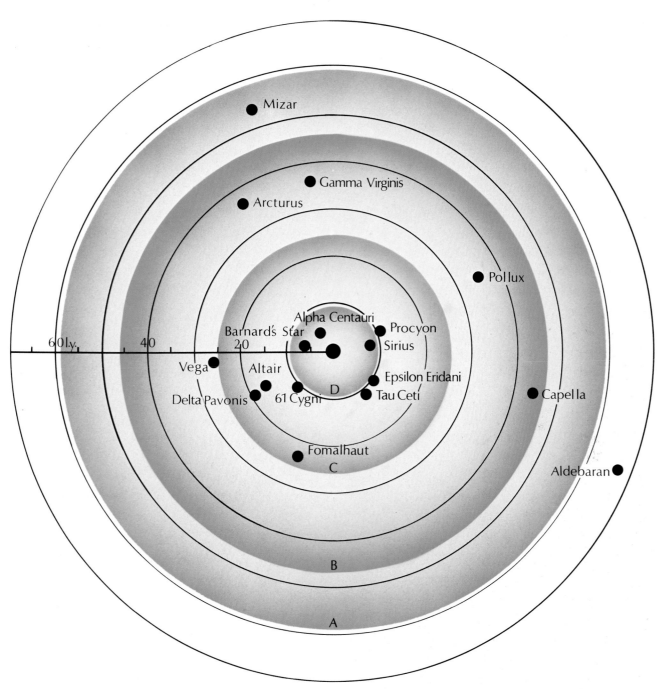

The distances in space reached by radio signals from Earth and the positions of nearby stars: A. 1923 – early days of broadcasting; B. 1936 – continuous radio broadcasts; C. 1957 – signals from Sputnik 1; D. 1974 – signals to possible extraterrestrial civilizations.

greater distances the Earth cannot yet be heard. From Aldebaran, the bright orange star in the constellation of the Bull, our Super Telescope would show us the Earth as it used to be 69 years ago; this takes us to 1914, and the start of World War I, which had so tragic and so long-lasting an effect on human civilization. Aldebaran is the first red giant star we have encountered in our travels. It has a diameter of 35,000,000 miles, and is at least 100 times as luminous as the Sun, but it is much less dense; its outer layers are much more tenuous than the air you and I breathe. It has also used up its store of available hydrogen 'fuel', and has started to draw upon its reserves.

If we are approaching Aldebaran from the direction of the Earth, something else is worth noting. In the background, several tens of light-years further away, we can see a whole collection of stars making up a true cluster, that of the Hyades. From Earth it looks as though Aldebaran is immersed in the cluster, but this is not so. Aldebaran lies about half-way between the Hyades and ourselves. There are no three-dimensional effects in space, and appearances can be highly misleading.

Now that we have travelled so far from home, the Sun has been lost in the starry sky; it is merely one of many millions, and there is nothing to mark it out from any other Main Sequence yellow dwarf. But before making a prolonged stop at our next important target, Betelgeux in Orion, we must pause to look at two other stars, Alkaid in the Great Bear and Nunki in the constellation of Sagittarius, the Archer.

From Earth, Alkaid and Mizar appear side by side; they are the first and second stars in the Great Bear's tail. There is an obvious temptation to think of them as being associated, but once more appearances are deceptive. Mizar, as we have found, is 59 light-years away; Alkaid is 108 (rather closer than was thought until recently; earlier measures put its distance at as great as 210 light-years). The point is that Alkaid is nearly as far away from Mizar as we are. On our light-beam we can pass first Mizar and then Alkaid; once Mizar has been left behind, it and Alkaid will appear on opposite sides of the sky, which is an added reminder that what we call a 'constellation' has no real significance. Indeed, from Alkaid the star patterns will take on a completely unfamiliar aspect, and the familiar Bear, Orion, Scorpion and the rest will be quite unrecognizable. Neither would Sirius appear particularly bright, now that we are seeing it from a distance of over 70 light-years.

Use our Super Telescope from Alkaid, and we would see the Earth as it used to be in 1875. In the United States, General Ulysses S. Grant was President; the Wild West was still the Wild West with its cowboys, though the most turbulent period was over; memories of the Civil War were fresh in men's minds. In the Middle East, the Suez Canal had yet to be built. In England, Queen Victoria was on the throne, while on the cricket fields in Gloucestershire (and elsewhere) W.G. Grace was scoring century after century. Moving out another 100 light-years we come to Nunki, a really powerful star with over 500 times the luminosity of the Sun. It is hot and bluish white; it may or may not have planets (the prospects do not seem good), but it would allow our Super Telescope to show us the Earth in 1774, so that we could watch the American Declaration of Independence being drawn up.

On now to Betelgeux, one of the most famous stars in the sky. Few people can fail to recognize Orion, the celestial Hunter, who dominates the evening sky during winter; the two brightest stars are different, since Betelgeux is orange-red, Rigel pure white. The pattern of the Hunter is distinctive, with three bright stars forming the Belt, and the misty Sword below. But we must make one admission. Betelgeux is several hundred light-years away, and our parallax method, which worked so well for the closer stars, has broken down simply because the shifts have become too small to be measured with any accuracy; they are swamped in unavoidable errors of observation. We have to turn to other methods, most of which involve studying a star's spectrum and deciding how luminous it really is, in

Clouds in space; we have to circumnavigate them – this one is in the Scorpion, and contains both Antares and the star Rho Ophiuchus. The cluster, M.4, 5700 light-years from Earth.

which case its distance can be calculated, provided that we allow for complications such as the absorption of light in space. The precision obtainable is obviously less. Betelgeux has been said to be about 520 light-years away. The most recent estimates reduce this to about 310 light-years, in which case the Super Telescope will show us the England of King Charles II, and we can watch the great scientists of the time as they go about their work – the Reverend John Flamsteed, to become the first Astronomer Royal at the observatory set up in Greenwich Park by express order of the King; Edmond Halley, of comet fame; and the greatest of them all, Sir Isaac Newton. Unfortunately, we have to accept comparatively large uncertainties with star distances beyond 300 light-years or so. All we can say is that they are certainly of the right order.

So let us approach Betelgeux, a true stellar giant. It is relatively cool, with a surface temperature of a little over 3,000 degrees centigrade, which is why it is orange-red; but to make up for this, it is huge. The diameter is always at least 450,000,000 miles, and may swell out to considerably more, because Betelgeux is a variable star; it is unstable, and both its size and its energy output change, though not very quickly. From Earth, Betelgeux may sometimes be almost as brilliant as the other leader of Orion, Rigel. At other times it may be no brighter than Aldebaran. The mass is equal to about twenty Suns, and the luminosity would match 15,000 Suns.

In fact, the mass is not so great as might be expected for so vast a star. Much of Betelgeux is made up of very rarefied gas, corresponding to what we would normally call a laboratory vacuum. But deep inside, conditions are remarkable. Betelgeux is old; it has exhausted its available hydrogen fuel and has started to use other nuclear reactions to produce its energy, building up heavier and heavier elements in the process. The temperature at the core is incredibly high, as is true of all red giants. When all its reserves are exhausted, Betelgeux will not shrink quietly into the white dwarf condition, as the Sun will. It is much more likely to explode with catastrophic violence, in an outburst of the type we call a supernova.

Let us look back at the H-R Diagram for a moment. Betelgeux must have begun by condensing out of interstellar material, and we can then place it to the right-hand side of the Main Sequence. Gradually it shrank, hotted up, and joined the Main Sequence near the upper left. But it is a rule in the universe that massive stars are spendthrifts; they use up their fuel far more quickly than modest stars such as the Sun, and so Betelgeux remained on the Main Sequence only briefly by solar standards. It then swelled out as different reactions began inside it, and turned into the red giant of today, to the upper right of the H-R Diagram.

Because it is so large, and because it is one of the closest of the red giants, modern-type equipment has enabled us to record some patches on its surface, though of course ordinary telescopes still show it as nothing more than a point. Certainly there are darkish areas, perhaps tremendous 'starspots', and we cannot doubt that there is ceaseless activity; Betelgeux is coming to the end of its brilliant career, and it has started to show the eccentricity of old age. Any planets which may once have attended it will have been swallowed up as the star swelled from its Main Sequence size to its giant phase unless, of course, they were sufficiently far out; not millions of miles, but thousands of millions of miles. Such a planet would be lit by a baleful red glow, and the edge of Betelgeux would be blurred and ill-defined instead of clear-cut like that of the Sun, because the outer layers of Betelgeux are so tenuous.

In our imagination, we can picture the plight of a civilization on a planet moving round Betelgeux at such a distance. The temperature now could be tolerable; but when Betelgeux collapses, the planet will be seared in a colossal blast of radiation, to be followed by numbing cold as the present Betelgeux changes into something small, super-dense and very feeble. Unless the inhabitants of the planet have been able to master the secrets of interstellar travel, there can be no hope for them.

Slightly further away from the Solar System, so that our Super Telescope will show us what is now the United States in its very early colonial days and England in Commonwealth times, there is another star group worthy of our attention: Algol, the Winking Demon. Normally it is reasonably bright (about equal to the Pole Star), but

Impression of the Albireo system. The scene on this planet is gloriously illuminated by the contrasting blue and gold colours of the two suns of Albireo, in the constellation of Cygnus the Swan.

in the seventeenth century astronomers noticed that every two and a half days it 'winked', fading slowly for a few hours and remaining dim for only about twenty minutes before recovering. In 1783 the reason for this odd behaviour was discovered by a most unusual astronomer, John Goodricke. He was deaf and dumb, and died when only 21, but there was nothing the matter with his brain, and he realized that instead of being truly variable, as Betelgeux is, Algol is an 'eclipsing binary'. There are

two components, one about as powerful as the Sun and the other 100 times brighter; they move round their 'balancing point' in two and a half days, and they are so placed that at each revolution the fainter component passes in front of the brighter, cutting out part of its light and causing the 'wink'. The two Algols are only about 6,000,000 miles apart, so they cannot be seen separately from Earth, but if our light-beam spacecraft approaches the system from 'above' or 'below', there will be no variation in light. There is also a third member of the group, moving round the main pair in a period of less than two years at a distance of 50,000,000 miles, and a fourth component

The Pole Star is shown in the middle of the photograph. This is a time-exposure: the stars seem to move round the pole of the sky. The Pole Star, within one degree of the pole, produces only a very short tail.

has been suspected. Evidently we have found another stellar family, but again we can hardly expect a planetary system, so it is time to move on.

By now our Super Telescope is showing us events on Earth which happened a very long time ago. If we visit Albireo in the constellation of Cygnus the Swan, we can look back and see England as it was in Shakespeare's time. Moreover, the scene on a planet of Albireo would be magnificent, because the yellow, highly luminous primary star has a blue companion at a distance of some 400,000 million miles. This is great enough to allow for a system of planets round either star, though perhaps the blue member is the more promising, since it is 'only' 120 times as luminous as the Sun. On a planet of the blue Albireo the scenery would be variously and gloriously coloured by the light of the two contrasting suns, one golden yellow and remote, the other blue and close.

Antares in the Scorpion, at about the same distance from the Earth, is a red giant similar to Betelgeux, and is attended by a greenish companion star about 47,000 million miles from it. Antares may be larger than Betelgeux, in which case it could swallow up the whole of the inner part of the Solar System. But more imposing is Sheliak or Beta Lyræ, whose distance is rather uncertain but is certainly over 500 light-years.

Here again we have an eclipsing binary, but it differs from Algol inasmuch as the two components are not very unequal. Both are egg shaped; the larger of the two has a diameter of over 16,000,000 miles, and they are so close together that they almost touch. Their surfaces are less than 10,000,000 miles apart, so that each component is tidally distorted (hence the egg-like shapes). The more massive component, which is the fainter of the two, is pulling huge streamers of gas out of its companion, while there is also a tremendous cloud of gas enveloping the whole system.

We have here a very curious state of affairs. At present the bright primary has a mass eleven times that of the Sun; the fainter secondary has twenty times the solar mass. But as the secondary pulls material away from the primary, the masses are becoming less unequal, and there will finally come a time when the situation is reversed, with the primary outweighing its companion. As both stars age, they will explode, and the system in its present state will be destroyed. Even today Beta Lyræ is one of the most unstable systems we know, though it is only one of many other pairs of the same general type.

Certainly no planet could exist close to these strange stars. However, there is another member of the group 1,000,000 million miles away, also a close binary, with a total luminosity equal to 80 Suns. If this star has a planet, its sky must be unrivalled, with coloured streamers and quick changes in Beta Lyræ itself together with the radiance sent down by its own double sun.

It is quite refreshing to continue our journey to visit the Pole Star or Polaris, at 680 light-years, so that our Super Telescope will show us England in the time of Edward I. Wales and Scotland have been subdued, and the English Crown seems strong and secure, though the warrior king is to die shortly afterwards, opening the way for Robert the Bruce's campaigns against a much weaker king, Edward II.

Polaris is important to us because it lies almost exactly at the north pole of the sky as seen from Earth, so that it remains motionless with the entire heavens revolving round it once a day. Of course, this has nothing to do with Polaris itself, and in any case the Earth's axis shows a slight but steady shift in direction, so that the pole of the sky moves too; by AD 12000 Vega will be the north polar star. Polaris itself is easy to recognize. It is not outstandingly bright in our skies, but this is because it is so remote;

it is in fact very powerful, equalling 6,000 Suns, and its surface is somewhat hotter than that of the Sun.

We are now coming to a region in which only the highly luminous stars are conspicuous as seen from Earth; stars no more luminous than the Sun are lost in the general background. We come across red giants, supergiants and brilliant bluish stars; at just over 900 light-years we approach Rigel, which is certainly worth a visit. It is the leader of Orion, and is much more powerful and remote than Betelgeux, so that it would take 60,000 Suns to match it. Its surface temperature is 12,000 degrees centigrade, and its spectrum is of Type B. Rigel is very large, with a diameter of 40,000,000 miles or so, and its mass is 50 times the Sun's. If Rigel were as close to us as Sirius, it would send us as much light as the half moon.

Our uncertainties about distance measurements are bound to increase as we move further and further away from the Solar System. If Rigel is indeed 900 light-years away (and it can hardly be closer than that), our Super Telescope would show us the Earth as it used to be in Norman times, with the armies of William the Conqueror still busily subduing the Saxons. If Rigel is slightly more remote, we might well witness the Battle of Hastings, or even look in at the court of King Canute (who, despite the famous legend, was much too wise a monarch to sit by the sea-shore and order the waves to retreat).

Rigel is using up its reserves at a furious rate, and is likely to leave the Main Sequence within the next 400,000,000 years. If it possesses planets, they can hardly be suited to advanced life-forms. There is also a companion star, about 250,000 million miles from Rigel itself, which is dwarfed by its brilliant primary, but is itself quite powerful; it is 150 times as luminous as the Sun, which means that it easily outshines near-by stars such as Sirius, Vega and

Mizar. To be more precise, Rigel's companion is yet another close binary, but the components are only 2,000,000 miles apart and take a mere ten days to complete one revolution round their common centre of gravity.

Where next? We have a wide choice, but we would be wrong not to visit Canopus, which shines as the most brilliant star in the skies of Earth with the single exception of Sirius, though unfortunately it lies too far south to be seen from England. It is just visible from Alexandria, where it rises briefly above the horizon once every 24 hours (often in daylight, of course), but it is never visible from Athens. It was this behaviour which gave the ancient Greeks an early proof that the Earth is a globe rather than a flat plane, as was originally thought. Athens and Alexandria were two of the greatest centres of culture in classical times. If the Earth had been flat, then the visibility of Canopus from Alexandria and its non-visibility from Athens would have been impossible to explain.

Canopus has a surface slightly hotter than that of the Sun. It has often been described as yellow, though most observers will call it white. Estimates of its distance and power have shown wild discrepancies. One measurement gave its distance as only 120 light-years, but the most recent estimates confirm earlier views that it is extremely remote – and it is a true cosmical searchlight. If its distance is 1,100 to 1,200 light-years, it must shine with the power of at least 200,000 Suns, and from it our Super Telescope would show scenes of conflict in Britain; the Danish raids have started, and the Saxons are very much on the defensive. With good timing, we might even see the burning and sacking of Lindisfarne.

Impression of Beta Lyræ. The unusual binary system of Beta Lyræ is clearly seen from a nearby planet. The two components of the system are elongated by tidal distortion, while streamers of gas torn from the stars spiral out into space.

The Small Cloud of Magellan. It contains Cepheid variables, which may be regarded as equally distant from us; and because the longer-period variables looked the brighter, it followed that they really were the more luminous.

If Rigel and its kind are spendthrifts, then Canopus is even more so. With its tremendous mass and its unbelievable flow of energy, it cannot last in its present form for more than a few million years. If we linger for long enough, we will see disaster overtake it; but as we would probably have to wait for a good many centuries we will move out still further, and call in on one of the most famous of all stars: Delta Cephei.

Delta Cephei is not far from the north pole of the sky, but it is not striking, and has no individual name. It is even more remote than Canopus; 1,300 light-years is a reasonable estimate, so that from it we could use our Super Telescope to focus on Jarrow and see the Venerable Bede hard at work writing his books. What makes Delta Cephei so important is that it is a very special kind of variable star. It is not like Betelgeux, whose fluctuations are unpredictable and relatively slight; neither is it like Algol or Beta Lyræ, which are not genuinely variable at all. Delta Cephei brightens and fades with clockwork regularity, reaching its maximum brightness every 5·3 days. It is utterly predictable, so that we always know how bright it will look at any particular moment. It is pulsating; that is to say, alternately swelling and shrinking, changing its energy output in the process. Its average diameter may be as much as 26,000,000 miles, giving a mean luminosity of 6,000 Suns, which is about the same as the Pole Star.

Canopus, a celestial searchlight. Except for Sirius, it shines as the most brilliant star in the skies of the Earth. We pass it at a respectful distance: we cannot dare approach a star which shines about 200,000 times as powerfully as the Sun.

Delta Cephei is not unique. Many more members of the class are known, and have been called Cepheids after their prototype. The periods range from a few days up to several weeks, but all are in states of regular pulsation. They seem to be well advanced in their life-stories, and it has even been suggested that many stars pass through a Cepheid stage during their later careers.

Just before World War I, the American astronomer Henrietta Leavitt was studying Cepheid variables in a system known as the Small Cloud of Magellan. This is now known to be an independent galaxy, well outside the Milky Way. Miss Leavitt did not realize this at the time, but she reasoned that all the stars in the Cloud must be at roughly the same distance from us, just as for most purposes it is good enough to say that Oldham and Blackburn are the same distance from New York. Miss Leavitt found that there was a relationship between a Cepheid's period and its brightness; the longer the period between one maximum and the next, the brighter the star. Now, if all these Cepheids were equally distant, it followed that the longer-period stars were genuinely the more luminous.

It would be difficult to over estimate the importance of this discovery. Once a Cepheid had been watched for a few days, and its period found, its real luminosity could be deduced – and, hence, its distance from us. Cepheids act as 'standard candles'

NGC 4535, a barred spiral galaxy. The arms issue from a kind of 'bar' through the system giving rather the impression of a letter S!

in space, and because they are so powerful they can be seen over an immense range.

Travelling still further, we come to another Cepheid, Eta Aquilæ in the Eagle. Here the period is 7.2 days, and so it is more luminous than the shorter-period Delta Cephei itself. In the skies of Earth they appear almost exactly equal; the greater power of Eta Aquilæ means, then, that it is further away.

Climates on a planet orbiting a Cepheid variable would show a great range of temperature, and it is difficult to believe that advanced life could develop under such conditions. We must be thankful that our own Sun is a placid, stable star rather than a pulsating Cepheid!

Close to Delta Cephei in the sky, though not in space, is Mu Cephei, which the great observer William Herschel nicknamed the 'Garnet Star' because of its redness. In size and power it is probably superior to Betelgeux, but its distance is around 1,600 light-years, so that our Super Telescope would see England during the last stages of the Roman Occupation, with the legions just starting their final evacuation of our islands. And before we enter the next phase of our journey, there are two more stars which we should examine. One is Deneb in the Swan; to us it shines only as the nineteenth brightest star in the sky, but in power it is of the very first rank, inferior only to Canopus among the stars that we have already visited. It equals perhaps 70,000 Suns, and it is over 50,000,000 miles in diameter. Like Canopus, it is white, but its surface is slightly hotter, at a temperature of some 9,700 degrees centigrade. Our Super Telescope would show the Earth near the end of the time of real Roman greatness, with Antoninus Pius in power and the legions still firmly established in much of England.

We have been making liberal use of our Super Telescope; but from Deneb what would we really be able to see with equipment of the kind we have actually built? Of course, the Sun would be detectable, but only as a very dim star of the thirteenth magnitude. All the planets, even Jupiter, would be hopelessly beyond range. But let me stress, again, that we have been using our imagination. Theoretically, Roman Britain could be seen from Deneb. In practice, of course, it could not.

The last star we plan to visit before coming on to the more bizarre objects of the sky is yet another cosmical searchlight, Lesath in the constellation of the Scorpion. It is fairly bright in our skies, but not outstanding; it cannot rival Rigel or Deneb, let alone Canopus, but it is 18,000 times as luminous as the Sun. We are to visit it because from its distance, 3,500 light-years, we would see the Earth as it used to be around 1500 BC. This was the time of the greatest volcanic outbreak in near-historical times. In the Mediterranean, the island of Thera, or Santorini, blew up. The island itself was shattered; the blast was much more powerful than the Krakatoa outbreak of 1883, for which we have eyewitness accounts. Thera's explosion was a disaster. All life in the region was wiped out, and huge waves rolled across the sea, hurling themselves against the island of Crete and, according to some authorities (though not all), destroying the brilliant civilization which had developed there, known to us as the Minoan after the legendary Cretan King Minos. Go to Thera today, and you will find a quiet semicircle of islands, some of them inhabited. It is difficult to picture the fury which was unleashed so long ago. Nobody who was in the danger-zone survived to tell the tale, but if we could go to Lesath and use our Super Telescope we could see the whole tragic story being enacted.

Three thousand five hundred light-years . . . but even Lesath is not far from us on the scale of the cosmos as a whole. We will find this out when we pass on to the external galaxies or 'island universes'. But first there is more to see in our own Galaxy; there are objects which we must on no account miss during our journey.

7 Stranger than Fiction

Now that we have travelled so far on our light-beam, it is time to pause and consider the situation. We have reached a distance of 2,000 light-years, so that our Super Tele-scope will show us the Earth as it used to be at the time of Christ's birth. It would be fascinating indeed to look at the Holy Land, and to see whether there is any truth in the legend of the Star of Bethlehem – which was certainly not due to a comet, a supernova, a planetary conjunction or anything else which we can pin down scientifically. Meanwhile, we have visited stars of many kinds, from the brilliant to the feeble, from red to blue; we have seen star-pairs, star-families and stars which swell and shrink. But the Galaxy contains even stranger objects, some of them bizarre by any standards. It so happens that none of these cosmical oddities can be found in our own neighbourhood, but this is probably a good thing, because some of them are danger-ously violent.

We are coming up to something new. From a distance it looks like a shining cycle-tyre, but as we draw in we see it in its true guise; a small, hot, bluish star surrounded by a shell of gas. It is Messier 57, better known as the Ring, and the most famous of the objects which we call planetary nebulæ.

Actually the name is not a good one, because M.57 (so called because it was the 57th entry in a list of such objects compiled 200 years ago by the French astronomer Charles Messier) is neither a planet nor a true nebula. It represents a late stage in the life of a star. As we have found, a star such as the Sun is forced to change its structure when it has used up its fuel reserves, and in many cases its outer layers are thrown off altogether, producing a shell which ex-pands, becoming larger, fainter and more tenuous all the time. The Ring is in this condition now. The gas is so thin that we need not fear entering it; it is millions of times less dense than the Earth's air, and we can have a good view of the central star, which is still very hot even though pitifully shrunken. The planetary nebula stage will not be permanent. After a relatively brief period of a few tens of millions of years at most, the shell will have become so spread-out and so thin that it will cease to shine; the material will merge into the incredibly ten-uous medium of interstellar space. At the moment the diameter of the Ring Nebula is roughly half a light-year, and the central star still has a surface temperature of 10,000 degrees centigrade, so that it clings on to some of the departed glory of its giant stage.

The Ring is not unique, but it is particu-larly easy to locate from Earth, because it lies midway between two naked-eye stars, Sulaphat, or Gamma Lyræ, to one side of it and the remarkable eclipsing binary Beta Lyræ to the other. Of course, we are again dealing with a line-of-sight effect. The Ring is much more remote than either of the stars flanking it.

The Galaxy contains over 100 known planetary nebulæ, not all of them so regular

The Crab Nebula. As we approach this supernova remnant, we are aware of tremendous disturbances in the expanding gas; inside is the 'power-house', now known to be a pulsar.

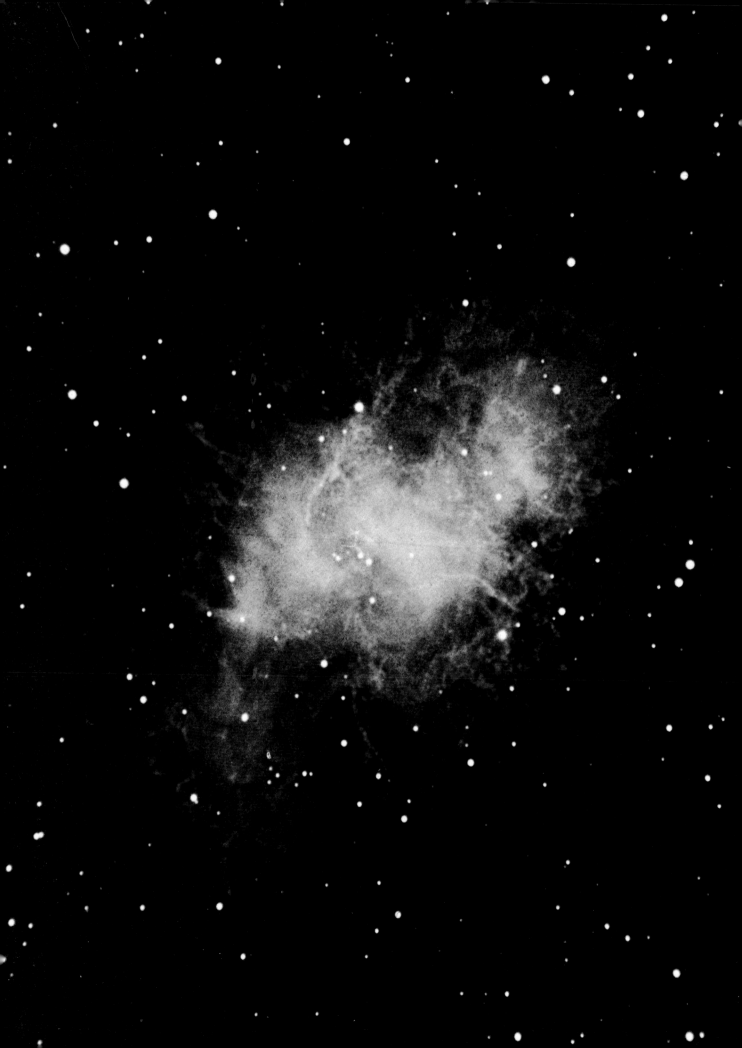

as the Ring. For example, there is M.27, in the constellation of the Fox, which has been nicknamed the Dumbbell Nebula for obvious reasons. M.97, in the Great Bear, has two stars inside its shell, giving it a strange resemblance to an owl's face. Not surprisingly, it is always called the Owl Nebula, but we are not yet within range of it on our journey; its distance from Earth is of the order of 10,000 light-years.

The Dumbbell Nebula in the Fox. We must not omit to look at this remarkable object during our journey. The Dumbbell (M.27) is a planetary nebula, but, unlike the Ring Nebula in Lyra, it is not symmetrical; its nickname is rather obvious.

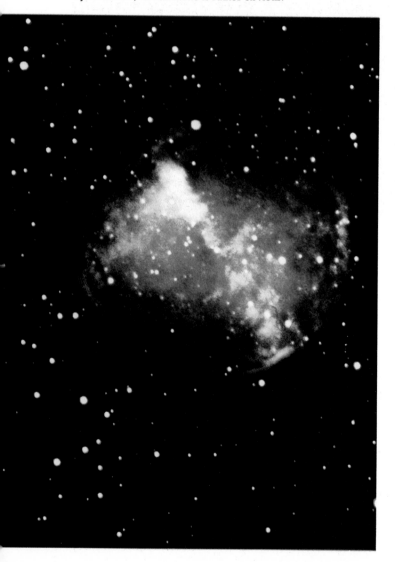

It is time to move on. Our next stop will be of special interest, because we are to go to one of the oddest objects in the entire Galaxy: the system of Epsilon Aurigæ. We have a long way to travel, because Epsilon Aurigæ is thought to be about 4,500 light-years away, so that our Super Telescope will show us the Old Kingdom of Egypt – perhaps even the building of the first pyramids – though again we cannot be precise simply because we cannot give really accurate figures for the distances of objects so remote as Epsilon Aurigæ.

Let us prime ourselves with a few facts before arriving. With the naked eye, Epsilon Aurigæ looks like a very ordinary star, one of a small triangle close to Capella in the sky; the three are known collectively as the Hædi, or Kids, and Epsilon is generally the brightest of them. I say 'generally' because Epsilon fluctuates in light. Every 27 years it begins to fade, continuing to decrease for six months until reaching its minimum value, though it never falls below naked-eye visibility. It remains faint for almost exactly a year, and then takes another six months to recover its lost light. The variability was discovered by a German astronomer named Fritsch as long ago as 1821, but not until modern times was it appreciated how remarkable Epsilon Aurigæ is.

If it 'winks' regularly, we are presumably dealing with an eclipsing binary – after all, we have stars such as Algol and Beta Lyræ, though in these cases the winks occur every few days. If the eclipse of the bright component of Epsilon Aurigæ lasts for so long, then the eclipsing secondary body must be very large. And here we come to our first problem. The bright primary is exceptionally luminous, and must be at least 60,000 times as powerful as the Sun, probably more. We can calculate that it must have a mass equal to 35 Suns, while the eclipsing companion has twenty times the Sun's mass. Yet the companion is completely

The Ring Nebula, M.57 Lyrae. One of the first of our 'oddities'; a very old star which has thrown off its outer layers, so that what is left is now surrounded by an expanding gas-shell. The central star is clearly shown.

invisible. There is no way in which we can track it; it does not emit visible light, radio radiation or anything else. But for its regular passage in front of the bright primary, we would not know of its existence.

Even as we travel towards it, we still cannot detect the mysterious secondary. If it really is a star, it must be of immense size, with a diameter of at least 2,000 million miles, which is big enough to engulf the orbits of all the planets in the Solar System out to beyond Saturn. In this case the secondary would be very young, and still condensing out of dust and gas, so that it has not yet started to shine. However, the bright primary can still be seen even when the eclipse is total, and it is difficult to believe in a huge, invisible companion which is also transparent. Moreover, no star of this size could be stable in the presence of a real heavyweight such as Epsilon Aurigæ A.

Suppose, then, that the secondary has not yet become a star, and is nothing more than a disk-shaped mass – a kind of cosmic biscuit – moving in such a way that every 27 years it bisects the primary as seen from Earth. This is a better idea, but it is still unsatisfying, and on the whole it seems more likely that the secondary is a small, hot star which is surrounded by a vast shell of gas dense enough to be partly opaque.

If this is correct, we should start to see indications of the secondary as we close in, but the blue star itself remains obstinately

Impression of the mysterious object Epsilon Aurigæ. It is a brilliant yellow star with an invisible companion which every 27 years passes in front of it, partially cutting off its light. Its companion may be a sun surrounded by an envelope of gas.

invisible, hidden by the cocoon in which it has wrapped itself. Now, however, we do start to pick up a certain amount of ultra-violet radiation, which is characteristic of extremely hot stars. Epsilon Aurigæ A is yellowish, and several millions of miles across; it produces tremendous tidal disturbances in the shell of its companion, and the scene is one of chaos, with swirling clouds and streamers everywhere. No planet, Earthlike or otherwise, could survive under such conditions. There can be no living thing in the incredible system of Epsilon Aurigæ.

There is another star, perhaps as remote as Epsilon Aurigæ, which is worthy of attention – though it is only fair to admit that estimates of its distance are very uncertain indeed, and it could be a great deal closer. It is known as R Coronæ, and it, too, is variable, but in a different way from Epsilon Aurigæ. For long spells, sometimes amounting to years, it remains steady in light; then, without warning, it begins to fade, and there is a prolonged minimum before the star recovers, jerkily rather than steadily. From Earth it is on the fringe of naked-eye visibility when at its normal brightness, but when near minimum, a telescope of considerable size is needed to show it.

Let us take a closer look at R Coronæ just as it starts to fade. We see what looks like a somewhat yellowish, very luminous star being slowly but steadily veiled by dark clouds which look like soot. If we go close enough, we find that the clouds really are soot. R Coronæ contains much less hydrogen than normal stars, but there is an excess of carbon, and it is these carbon particles which periodically billow up through the extended, tenuous atmosphere and block

out a large part of the star's light. After a while the soot begins to disperse, blown away by the powerful radiation from below, so that R Coronæ resumes its normal appearance. It is not unique, though R Coronæ stars are comparatively rare.

On to 5,000 light-years from Earth; we use our Super Telescope to show us Egypt as it used to be when originally united by King Menes, who had a long and prosperous reign before being unfortunate enough to be killed by a hippopotamus. England, of course, is populated by scattered tribes, and of London there is as yet no trace. We are reaching the Rubicon between historic and prehistoric times.

We are coming up to an object which was seen from Earth as recently as 1975, but which had by then long ceased to exist in such a form. We know it as V.1500 Cygni, and it used to be an extremely faint star, well below the limit of most of our photographic catalogues. Suddenly, in the early hours of 29 August 1975, a Japanese astronomer named Osada saw that there was a newcomer in the constellation of the Swan. It could not be overlooked, because it was at least as bright as the Pole Star, and altered the entire look of that part of the sky, though it had certainly not been there on the previous evening.* It was a new star, or nova. Its glory was brief; it faded quickly, and within a week had dropped below naked-eye visibility, after which it returned to its former obscurity.

Note that what was seen in 1975 did not occur in 1975; it took place 5,000 years earlier, in the time of King Menes, which is a further reminder that our Earth-based view of the universe is always out of date.

* Osada discovered the star during the Japanese night, when it was still daylight over Europe. As soon as darkness fell, the star was independently discovered by many other observers. I discovered it soon after dusk; I think that I was about eightieth on the list, but at least I knew at once what it was.

However, from our convenient light-beam spacecraft we can watch to see just what happened. First, V.1500 is not a normal star, but a binary. Of its components, one is a normal star, rather reddish and not particularly luminous, while the other is a white dwarf, which means that it has run through much of its life-story and has become senile, with no fuel reserves left. Yet it is still very massive, and its gravitational pull is tearing material away from the primary, so that a stream of gas flows between the two. The gas does not fall straight on to the white dwarf, but spirals towards it, and gradually a ring of material is built up. Much of this material, naturally, is hydrogen.

As we watch, we notice that a change is taking place. The hydrogen is becoming denser and hotter as more of it is pulled away from the primary star, and suddenly the critical temperature of 10,000,000 degrees centigrade is reached. This is enough to spark off nuclear reactions, and the result is a tremendous outburst; at the peak of the explosion, the formerly dim system is emitting 500,000 times as much energy as the Sun. Of course, it cannot last, and the activity soon dies down, but it is spectacular while it lasts.

We can well picture the plight of any inhabited planet orbiting a system of this kind. At any reasonable distance, the planet will be either destroyed or else made sterile

Nova Cygni 1975. As we watch, a new star bursts forth – or, more accurately, a binary star system exchanging mass suffers a tremendous outburst, and flares up briefly to prominence before subsiding again.

The Veil Nebula. Near Deneb we see what appears to be the remains of a supernova – the Veil Nebula. It is still expanding outwards from the explosion centre; but we cannot track any associated pulsar.

by the sudden, tremendous blast. But remember that the white dwarf has presumably passed through a red giant stage, so that any planets would have met their fate earlier. Novæ are not really rare; some of them become much brighter than V.1500 did, and in 1918 a nova in the constellation of Aquila, the Eagle, rivalled Sirius for a few nights, though it has now become a very faint telescopic object.

A nova outburst does not destroy either component of the binary system, and we even know of a few 'recurrent novæ' which have been seen to explode more than once. There are, too, some much less energetic stars, known as SS Cygni or U Geminorum variables, which show mild outbreaks of similar type every few weeks. However,

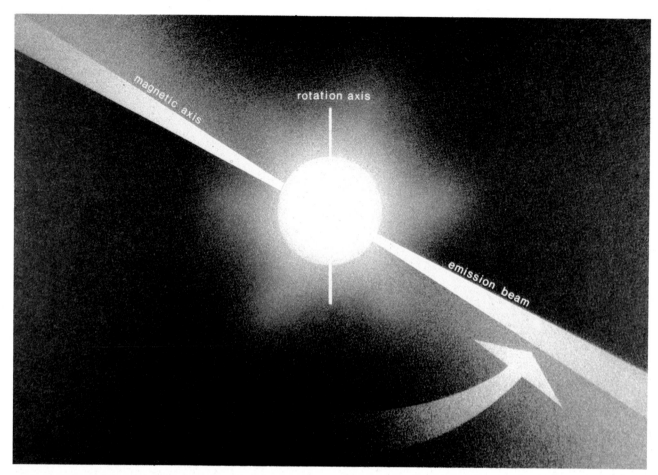

The 'Lighthouse' theory of pulsars. Radio waves are emitted by electrons at the poles of the magnetic field of the neutron star, or pulsar, and leave the star in the form of a narrow beam, sweeping around as the star rotates.

even V.1500 is gentle compared with the behaviour of our next target, the Crab Nebula, at some 6,000 light-years. From such a distance our Super Telescope will show the Earth barely recovering after the end of the last Ice Age, and we would see little of what we now call civilization.

The Crab Nebula (so called because a famous Irish astronomer of the last century, the Earl of Rosse, drew it with his powerful home-made telescope and commented that its shape was somewhat crab-like) is a huge cloud of gas. As we approach it, we can see that it is amazingly complex. There are streamers and filaments, and evidence of tremendous agitation. Moreover, the cloud is expanding in all directions, so that there must be some central 'power house' which is causing both the agitation and the expansion.

So far on our travels we have equipped ourselves only with our optical Super Telescope. Now let us conjure up an equally all-powerful radio telescope, and also an instrument for detecting very short waves, such as ultraviolet and X-rays. We find that the Crab is extremely powerful at all wavelengths. It covers the full range in a way that no object we have yet encountered has been able to do, so that the Crab is in a class of its own.

Diving deep into the turbulent gas, we find that the energy emitted becomes stronger and stronger. Then, at last, we see the cause: a tiny, rapidly spinning object, no more than a few miles in diameter, and rotating so quickly that it completes 30 turns per second. As it does so, it sends out pulses of radiation at radio wavelengths, together with flashes of light. We are being treated to our first view of a neutron star, or pulsar. It has been observed from Earth; it was first tracked down by radio astronomers and then, considerably later, by visual searchers.

The Crab Nebula is a stellar wreck. In the year 1054 Chinese and Japanese astronomers suddenly saw a brilliant star where nothing had been visible before; it became

bright enough to remain on view during daylight, and it lasted for several months before fading away. It was a supernova, the most violent explosion known in nature, and it has left us the Crab Nebula of today.

Theorists have managed to explain the sequence of events fairly well. We begin with a star considerably more massive than the Sun. It runs through the usual stages: initial contraction, Main Sequence and then red giant; but the great mass means that there can be no puffing-off of the outer shell to form a planetary nebula, followed by collapse into the white dwarf condition. Heavy elements are built up in the star's core, and finally the core itself is made up chiefly of iron. Now iron is a special element. It cannot continue the building-up process, no matter how high the temperature

Eta Carinæ. We must give this star a wide berth. It may have at least 6,000,000 times the power of the Sun, and it is not a normal star; it is associated with nebulosity, and it may soon (by cosmical standards) explode as a supernova.

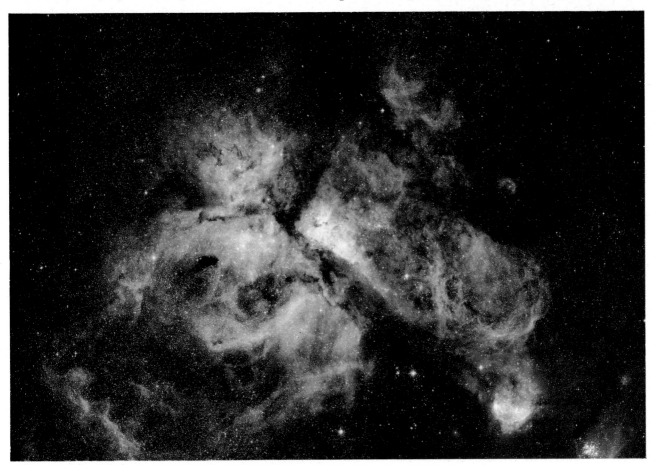

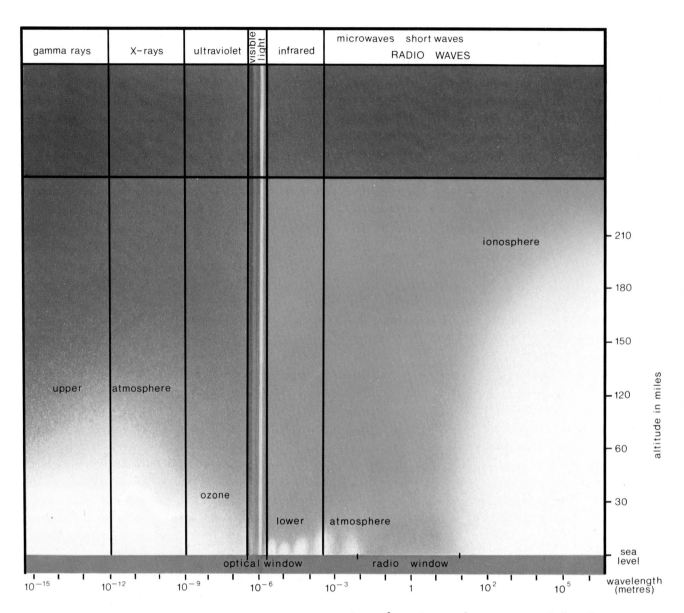

Our optical telescopes are limited when operating from Earth, because of the cut-off imposed by the atmosphere. Only certain wavelengths pass through. In space this restriction is removed.

– in this case, around 3,000 million degrees centigrade. There comes a moment when the core can no longer withstand the tremendous pressures bearing down on it. The iron atoms are broken up into their constituent parts, positively-charged protons, and negatively-charged electrons. Then, abruptly, events reach their climax. The protons and electrons are jammed together to make up particles known as neutrons, which have no electrical charge because the positive charges of the protons are balanced by the negative charges of the electrons. The whole star collapses in what may be termed an implosion. The shock-wave blows the star to pieces; the energy released is equal to at least 15,000,000 Suns, and the remnants of the star expand rapidly, leaving the neutron core behind.

The Crab neutron star, or pulsar, was not the first to be found; this distinction must go to a pulsar in the constellation of the Fox which was discovered in 1967 by Jocelyn Bell, a member of a team of radio astronomers at Cambridge. But the Fox pulsar was detected only by radio, and as yet the only two pulsars to be optically identified are those in the Crab and in the so-called Gum Nebula in Vela, in the southern sky. (The name does not indicate that the nebula is sticky. It honours an Australian astronomer, Colin Gum, whose brilliant career was tragically cut short by a skiing accident.) The Crab pulsar is probably the nearest known, but what is it like?

If we are to visit a neutron star, we must use our imagination even more lavishly than we have done up to now. Though the mass of the Crab pulsar is no greater than that of the Sun, the small diameter – perhaps six miles – indicates incredible density: something like 100,000,000 million times that of water. If we could withstand the tremendous pull of gravity (which, let it be said, no known material could do), we might find that the outer layer of the pulsar is made up of crystallized iron, while beneath come several layers made up chiefly of neutrons together, probably, with some even more fundamental particles about which we know nothing at all; we call them hyperons, but we are really doing no more than guess. It has been said that a pulsar is like an ancient raw egg: a solid shell, with several peculiar fluids inside. And there is, of course, an immensely powerful magnetic field.

Staying just clear of the pulsar, we find that the radio emissions are coming out only in two directions, on opposite sides of the star, rather like the beams of a rotating searchlight. Each time a beam sweeps over us we receive a pulse of energy, and it is this which produces the curious 'ticking' which so baffled the early investigators. For a few hectic days it was even thought possible that the signals were artificial, though it did not take long for this fascinating theory to be rejected. (At the time it was known as the LGM or Little Green Men theory. Prudently, the Cambridge astronomers did not announce it until they had made quite certain that it was not correct.)

All pulsars seem to be slowing down, admittedly by very small amounts, so that eventually they will cease to pulse. The Crab pulsar, which is the fastest-spinning of the normal pulsars, is therefore presumably the youngest. Others have rotation periods of several seconds, and it is possible, though not certain, that all are wrecks of supernovæ. Twentieth-century astronomers regret that there has been no supernova seen in our Galaxy since 1604, a few years before telescopes were invented. We have had to make do with the remnants, and with supernovæ that can be observed in other galaxies millions of light-years away.

We cannot tell when the next supernova will blaze out. It could be tomorrow; it might not be for centuries. However, there is one possible candidate, again about 6,000 light-years from us. It is Eta Carinæ, in the southern constellation of the Keel, unfortunately too far south to be seen from anywhere in Britain.

Eta Carinæ has an interesting history. A century and a half ago it shone as the brightest star in the sky apart from Sirius, but it slowly faded, until it had dropped below naked-eye visibility; today it is easily seen with binoculars. It is not a normal star. It is immersed in a gaseous nebula which was discovered 70 years ago and is steadily expanding; it is often called the Homunculus, because its shape reminds one slightly of a manikin. There is also a more extensive nebula, a whirling chaos of gas and dust cut by dark lanes; here, too, is the Keyhole Nebula, which seems to be expanding outwards at 25 miles per second.

Approach Eta Carinæ, and you will be struck by its colour; even Earth-based telescopes can show that it is like a red blob.

Impression of Cygnus X-1. The invisible companion of this binary star may well be a black hole which is pulling matter from the large blue supergiant star. As the matter spirals into the black hole, X-rays are emitted.

Eta Carinæ is hiding itself. When at its brightest, as seen from Earth in the 1830s, it must have been radiating about 6,000,000 times as powerfully as the Sun, and there is no reason to doubt that it is any the less luminous today, though much of its light is blocked by the Homunculus. If these estimates are correct (and they are founded upon very good evidence), then Eta Carinæ is the most luminous star in the whole of the Galaxy.

There are other puzzling features. If we add yet another piece of equipment to our armoury, and measure the surface temperature of Eta, we find it to be 25,000 degrees centigrade. This is cooler than might be expected. The diameter may be of the order of 8,000,000 to 9,000,000 miles, which again is on the small size when we remember giants such as Betelgeux. Where Eta Carinæ differs from other stars is in its mass, which may be as much as 100 times that of the Sun.

Now, there is a limit to the mass which a star can have; above this limit it becomes unstable. Eta Carinæ is probably close to the danger zone. It can emit shells of material, and has been caught in the act, but this may be doing no more than delay the inevitable. Eta Carinæ may be on the verge of 'going supernova'. If it does, it will shine in our skies more brightly than Venus while the outburst lasts.

The explosion may not come yet; it may be delayed for hundreds, thousands or even a few million years, but not for longer than that, assuming that our present ideas are correct (as is believed by most astronomers, though not all). Obviously, the outburst would devastate every object over a wide area before fading down to become a Crab-type gas cloud with a Crab-type pulsar. Of one thing we may be fairly confident. If we could return to the scene in, say, 100,000,000 years from now, the Sun will not have altered much, but the Eta Carinæ which we know will no longer exist.

Only massive stars can explode as super-novæ. There are no such stars within range of the Solar System; Betelgeux, one of the nearest, is over 300 light-years away, and though it may eventually produce a super-nova it could do us no harm, though it would provide us with a magnificent display of cosmic fireworks. Supernovæ are certainly objects to be studied from a respectful distance.

Even now we have not come to the weirdest of all objects in the Galaxy. For this

we must travel still further, out to 8,000 light-years, so that our Super Telescope shows us the Earth near the end of the last Ice Age, with mammoths, sabre-toothed tigers and other creatures which we know only in museums. Our destination is a giant star, referred to by its catalogue number of HDE 226868 but better known as Cygnus X-1. Again the distance is uncertain, but 8,000 light-years is probably not very wide of the mark. Even so, the star is visible from Earth with a small telescope. It is very massive (though by no means the equal of Eta Carinæ), and it is a very luminous supergiant with a diameter of over 10,000,000 miles. It is a member of a binary system, and we deduce that the companion has about half the mass of the visible star: that is, about fifteen Suns. As with Epsilon Aurigæ, the secondary is

The Southern Cross in the Milky Way. As we travel further and further from home, we see the Southern Cross in the Milky Way. The Eta Carinæ region lies to the right.

Now and then we see dark patches. These are simply nebulæ which are not being lit up. Barnard 66 (MGC 6520) is one; it may be a region in which stars are slowly forming.

invisible, but this time there is a difference. Cygnus X-1 is a powerful source of X-rays, which explains its name.

X-rays are produced by intensely heated material, and evidently there is something curious about the Cygnus X-1 system. Instead of a giant star accompanied by a pulsar, we may have encountered something even more extreme: a Black Hole.

Come back for a moment to the case of a very massive star which collapses when its energy supplies run out, producing a shock-wave and a supernova outburst. This can certainly happen. But if the original star is more massive still, the sequence of events may follow a completely different course.

Once the collapse starts, it is so sudden and so violent that nothing can stop it. The star goes on shrinking and shrinking, becoming denser and denser as it does so – and the escape velocity goes up.

We have come across escape velocity before; it is the speed necessary to escape from a body, seven miles per second for the Earth, 383 miles per second for the Sun, and so on. The smaller and denser the body, the greater the escape velocity. With our massive, collapsing star, there comes a point when the escape velocity reaches 186,000 miles per second. This, of course, is the velocity of light, and so not even light can escape from the shrunken star. The end product is an area that is to all intents and purposes cut off from the rest of the universe. It has become a Black Hole.

Theoreticians speculated about the possibilities of Black Holes long before the first spacecraft flew, but proof was difficult simply because a Black Hole emits no radiation of any kind, and therefore cannot be seen. If it is 'on its own', it will be completely undetectable. But if it is associated with a visible object there may be a chance, and the best candidate, so far, is Cygnus X-1.

Let us go in to take a careful look. HDE 226868 itself is blindingly brilliant, but it is not a normal star. Streams of gas are being torn out of it, and are making their way towards a region which at first sight appears blank. Then, as we close our range, we see that the gas and dust are spiralling round a definite centre, becoming so heated that it sends out a flood of X-rays. This, then, is the answer. The invisible secondary is a Black Hole. It can pull material away from its companion, as it is doing all the time, but from it absolutely nothing can escape.

The boundary beyond which all communication is cut off is called the event horizon. In our journey we can approach it; we see nothing as we pass through, because there is no solid barrier, but once we are inside everything changes, and all the ordinary laws of nature break down. In fact, any astronaut attempting such a trip would be instantly destroyed by the collapsed star's gravitational pull. Assuming that he passed over the event horizon feet first, the pull upon his legs would be thousands of times greater than the pull upon his head, because his head would be further away from the collapsed star, so that he would be promptly stretched out like a stick of spaghetti. Not that this would make much difference to him; he would have been doomed from the moment he entered the forbidden zone.

Impression of a Black Hole at the centre of a galaxy, possibly ours. The gravitational attraction of a Black Hole is so great that it pulls in all matter around it.

Assume that we can endure all these forces, and plunge inwards. Our time-scale is altered; strangely, an outside observer would see us 'frozen' at the event horizon, because on his reckoning we would take an

The Ring Nebula from close by. After a few tens of millions of years the shell will have so spread out that it will cease to shine. We may marvel at its beauty – but it represents a star in its extreme old age.

infinite time to pass over, but we could never signal back. Trying to see the collapsed star itself, we are baffled. There are suggestions that the star may crush itself out of existence completely when it has contracted so much that all its mass has been concentrated in a point of infinitely small size, known as a singularity. Exotic ideas have been proposed. If we avoid the actual

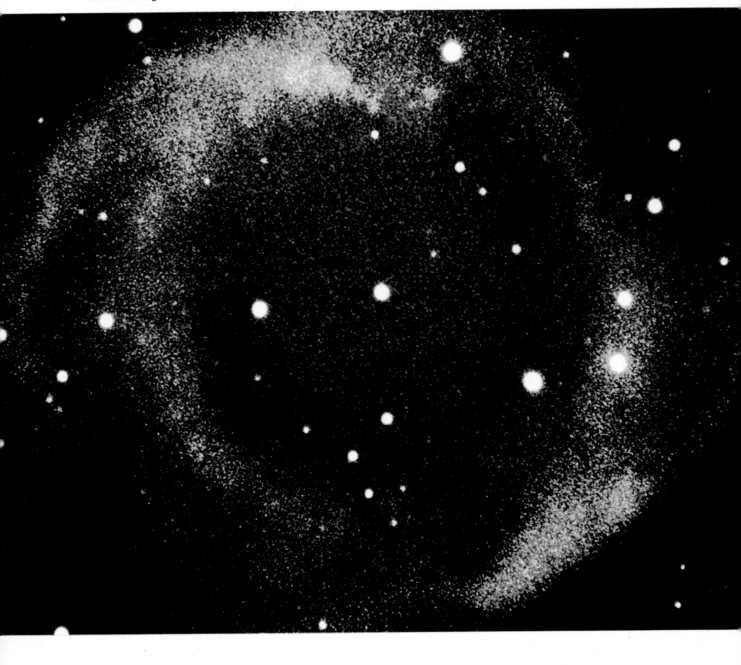

centre (as might be possible with a rotating Black Hole), could we emerge into a different part of our universe, or in a different universe altogether? There is no knowing, and at the moment there seems little chance of finding out.

In theory, a body of any size can become a Black Hole. Shrink down our Sun to a diameter of three and a half miles, and it will behave in just this way; for the Earth, the limiting size (known technically as the Schwarzschild radius, after the German astronomer Karl Schwarzschild) is one centimetre. But a Black Hole is greedy. It gobbles up material, and as its mass increases so does its size. Some writers have given gloomy pictures in which all the matter in the universe will be eventually swallowed by Black Holes. . . .

The plain truth is that we can do no more than make rather wild guesses, and if we start to visualize conditions in 'another universe', reached by way of a Black Hole, we have really entered the realm of science fiction. At any rate, there is no doubt that we have reached the ultimate in strangeness. There can be nothing more bizarre than a Black Hole.

Cygnus X-1 may be the best candidate, but there are others, and one of these, SS 433, is a good choice as a place to end our journey before we prepare ourselves for a dive into the very centre of the Galaxy. SS 433 is much more remote than Cygnus X-1; from it our Super Telescope would show the Earth deep in the last Ice Age. But we cannot miss this remarkable system, which may be aptly termed the Cosmic Lawn-Sprinkler.

Its story began, so far as we are concerned, when astronomers at Cambridge identified a supernova remnant, which was listed as W.50. Like Cygnus X-1, it sends out X-rays as well as radio waves, and inside it there seemed to be a faint, variable X-ray source. Further investigations showed that it was quite unlike any other known object.

Once again we have a binary system, and as we approach we can see that the primary is a luminous giant, while the other shows all the symptoms of a Black Hole. The Black Hole is, conventionally, pulling gas and material away from the giant, but so much gas is involved that even the Black Hole cannot digest all of it. Some of it is rejected, and streams out from either side in two opposite jets. The jets move at an amazing speed – at least a quarter that of light – and as they travel, they wobble and spray out in a cone, waving around in lawn-sprinkler manner. Moreover, the radio shell (W.50) is elongated, with two 'ears', which have evidently been blown out by the jets. It is an unstable system, and at the moment the only known example, though no doubt we will eventually find others.

Quite recently, a new theory has come to the fore. Quasars, which we will visit later in our journey, are remote and almost incredibly luminous. It may well be that a quasar is simply a galaxy with a Black Hole in its centre. If the Black Hole is small, it pulls in dust and gas, but very little radiation is given off by the material spiralling inwards. If the Black Hole is very large, it can swallow stars whole, so that again there is not much outward disturbance. But if the Black Hole is of medium size, it pulls its victim stars apart before it can swallow them – and this creates a tremendous disturbance, with radiation given off. If this picture is correct then, a powerful quasar is simply a galaxy with a medium-sized Black Hole in its core.

You will agree, I think, that the Galaxy is an intriguing place. In our journey so far we have been able to stop off only at a very small selection of objects; much has been left out. But our time is limited, and we must now make ready for the next trip, which will take us into the very heart of the Galaxy.

8 The Heart of the Galaxy

During our flight across the Galaxy we have seen a great many stars. Altogether, we can calculate that the grand total must be about 100,000 million. We have also noticed that in some areas the stars are more crowded than in others. What we have not been able to do is to gain any overall impression of the shape of the Galaxy, for the excellent reason that we are inside it. Of course, the patterns of stars have altered. Now that we are thousands of light-years away from our starting-point in the Solar System, the familiar constellations have become totally unrecognizable, and the stars that we have always regarded as being pre-eminent, such as Sirius, have faded into the distance.

Another point which we must bear in mind is that on our light-beam we have been able to move unhampered by any friction against the thinly spread material between the stars. Yet space is not empty; there are huge clouds of cold hydrogen, for example, and even though they are so tenuous (more so than the best vacuum we yet can create in an Earth laboratory) they make their presence felt by sending out radio waves at one particular wavelength, just over 21 centimetres. It was by plotting the positions of these clouds, during the 1950s, that astronomers were given the first definite proof that the Galaxy is a spiral.

But this can be discussed later. At the moment we have set course for the centre of the Galaxy itself, and we must aim for the lovely star clouds which, from Earth, are seen in the constellation of Sagittarius, the Archer. We are unquestionably travelling in the right direction, but we cannot see through to the galactic centre, because there is too much 'dust' in the way.

Dust, even very thin dust, is very efficient at absorbing light. (Gas, on the other hand, makes very little difference unless it is reasonably dense.) The star clouds of Sagittarius are dusty places, and our Super Telescope is a visual instrument, so that it cannot pierce them. Instead we must use our Super Radio Telescope. At once signals come through; something in the centre of the Galaxy is sending out powerful transmissions. There is infra-red radiation, too, and evidently we are approaching a region which is very curious indeed.

How far must we go? Measurements indicate that the centre of the Galaxy lies about 33,000 light-years from the Solar System, though some recent work has shown that this may be a slight over-estimate. In any case, radio observations tell us that the spiral arms rotate round the centre of the system in a 'trailing' manner, as the sparks sent out from a spinning Catherine-wheel firework will do. But when we have penetrated to about 9,000 light-years from the centre we find something else. There seems to be a spiral arm that is expanding outwards from the centre, and as we travel on we find other arms and star clouds, which are similarly expanding at velocities of up to 250,000 miles per hour.

Impression of the Trifid Nebula; as we approach it we see the gas clouds in all their glory, together with dark lanes which represent nebular material not lit up by suitable stars. The surrounding stars lies in the foreground.

Next we come to a vast disk of cool hydrogen which is not expanding, and we also find a ring of clouds which contain molecules.

A molecule is an atom group; thus a water molecule consists of two atoms of hydrogen together with one of oxygen, which explains the chemical formula H_2O. Carbon dioxide, which makes up so much of the atmospheres of Venus and Mars, consists of one atom of carbon and two oxygen atoms: formula CO_2. (It is also the gas that makes soda-water fizzy.) However, molecules are broken up by high temperatures, and are separated back into their constituent atoms. When the temperatures rise still further, the atoms themselves are disrupted into their constituent nuclei and electrons, and in such exotic objects as pulsars the nuclei are themselves jammed together. The centre of the Galaxy is presumably hot, and yet the clouds through which we are passing on our journey contain molecules.

NGC 6589. We are making our way, on our light-beam, towards the Sagittarius star clouds lying near the centre of the Galaxy; we come now to a region of dust and gas which we must study with our Super Radio Telescope.

The Milky Way with the trail of an Echo satellite stretching across it. (The two Echoes were balloon satellites of the 1960s).

Mixed up in the clouds we also find hydrogen atoms that have been disrupted by powerful radiation. The conventional picture of an atom is of a tiny Solar System, with a central nucleus representing the Sun and orbiting electrons representing the planets. The analogy is not a good one, because we cannot regard either protons or electrons as solid lumps, but it will serve us for the moment. In certain parts of the Galaxy we find what are termed H.II regions, in which the single orbiting electron in each hydrogen atom has been ripped away, leaving the atom incomplete or 'ionized'. It follows that an H.II region contains protons and electrons moving about independently.

Some H.II regions are near enough to be examined in detail; these are the gaseous-nebulæ, such as that in the Sword of Orion, which is easily visible with the naked eye on a clear night. Inside these nebulæ, fresh stars are being created out of the scattered material. One would hardly expect such processes to be going on so close to the centre of the Galaxy, and yet, apparently, it is happening.

Passing through the ring of clouds, which is around 1,200 light-years in diameter, we reach another cloud of hydrogen at a higher temperature. It is arc-shaped; indeed, astronomers call it the Arc. It, too, is moving outwards, at a rate of about 30 miles per second, and it can be classed as an H.II region. Still closer in there is a strong radio source, and as we finally emerge from the inner edge of the Arc we can make out details in the mysterious central region.

There are three main features. One of them, known as Sagittarius A East, is obviously a supernova remnant. Another, Sagittarius A West, seems to be surrounding the exact centre; as yet we cannot see what it contains. Between these two there is a huge cloud of molecules which is rushing outwards at nearly 25 miles per second. It is not a normal H.II region; there is no evidence of glowing hydrogen, so it does not look as though star formation is going on there. Exactly what it indicates is still not known. It could have been shot out from the centre in some cataclysmic outburst, in which case its speed shows that the outburst must have happened less than 1,000,000 years ago.

However, it is Sagittarius A West that interests us most. Travelling on, and penetrating it, we come next to a thin, spinning disk of gas which is, by cosmical standards, very small. It is less than seven light-years across, so that it would not even span the distance between our Sun and Sirius. It is rich in stars; we cannot see them from Earth, but we know that many of them are orange or orange-red, so that they are old. We can also detect sources that appear to be very young stars, still surrounded by their dusty shells, which have not yet been blown away. The whole region is intensely active, and the stars are relatively close to each other. Instead of being separated by light-years, as in the region of the Solar System, they are only light-days apart.

The chances of any life in this part of the Galaxy seem very slim. For one thing, there is an intense flood of radiation which would be lethal to any form of life we know. Secondly, the whole region is too active to allow for any leisurely process, such as the formation of a planetary system. Yet if there were any life-supporting planet, its inhabitants would have a superb view of the Galaxy's core.

Assume, then, that we find such a planet and land upon its surface. We are moving round an orange star, and we are close to it, so that the radiation is brilliant even though the star's surface is cooler than that of the Sun. Even in broad daylight we can see other stars, some of them rivalling the light of the Moon as seen from Earth. At night there would be brilliant stars everywhere, some orange-red, some blue and some white; many of them would show disks, casting immensely complex shadows across the planet's surface. Darkness would be unknown. What we call 'night' would be incomprehensible to a denizen of a planet in the central cluster.

How many stars? Plenty of them, far too many to count, but there are other ways of finding out. We can measure the total amount of radiation being emitted, and then do some averaging, because we can assume a mean luminosity for the stars; some will be brighter than the average and some fainter, but on the whole our estimate of 1,000,000 stars is probably not far wrong. This raises another problem. The disk is stable, and yet the combined mass of a mere 1,000,000 stars should not be enough to prevent it from dispersing. We need about 5,000,000, and we may be sure that there are fewer stars than that. We have to explain the missing mass, and the key must be sought in the very heart of the cluster, where there is a small, compact radio source.

Impression of the view from a planet moving round an orange star in the region not far from the galactic centre. The sky is filled with stars of all colours and a huge disk of stars surrounded by gas and dust marks the heart of the galaxy.

Before going right in, we may pause to consider the possibilities. We know that the ultra-compact region is tiny; its diameter cannot be more than about 1,000 million miles, about the same as that of the orbit of Jupiter, and it may be even less. What we have to do is to find something that can produce so much energy in so small an area.

Chains of supernovæ, perhaps? A single supernova can reach over 15,000,000 times the luminosity of the Sun, and several combined might be enough to power Sagittarius A West. But on reflection, the theory seems to be untenable. There is no reason to believe that one supernova outburst will trigger off another; moreover, no supernova stays at peak power for long, and Sagittarius A West must be thousands of millions of years old.

Another suggestion involves what may be termed anti-matter, the exact opposite of ordinary matter. This idea was developed by the Swedish astronomer Hannes Alfvén, who believes that whole galaxies of anti-matter exist. From afar they would look just the same as galaxies of our kind of matter; but if an ordinary atom and an anti-matter atom meet, they annihilate each other and leave nothing but a flash of energy. Certainly the energy produced would be considerable, but again we are up against

Journey to the centre of the Galaxy. Taken in stages, as though we were looking through a microscope, we see (*below*) 1 the spiral arms encountered on the way to the nucleus; (*and right*) 2 the nucleus of the Galaxy itself; 3 features inside the molecular ring; 4 Sagittarius A West and East and 5 a rotating disk of gas at the core of the Galaxy, showing the radio source.

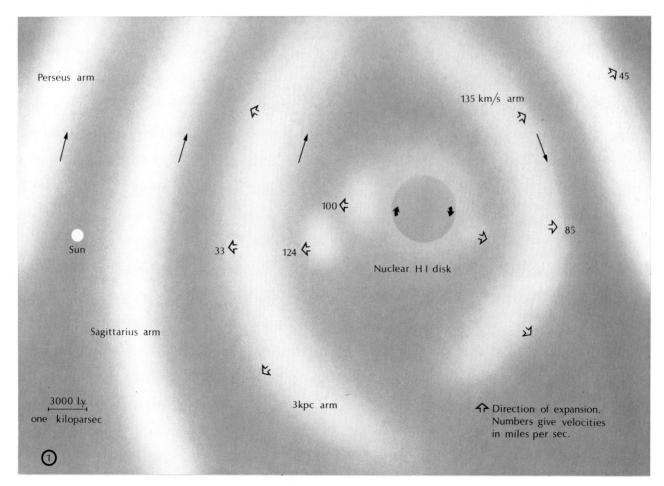

Perseus arm

135 km/s arm

45

100

Sun

33   124

Nuclear H I disk

85

Sagittarius arm

3000 l.y.
one kiloparsec

3kpc arm

↗ Direction of expansion.
Numbers give velocities
in miles per sec.

①

the time scale. If the process were operating, all the matter in Sagittarius A West would have been annihilated long ago. Moreover, there is absolutely no evidence that Alfvén's type of anti-matter exists in Sagittarius A West or anywhere else.

Gravitational energy is no more promising, and the time scale also rules out the idea of constant stellar collisions, which in any case could never generate as much energy as we need. No: there seems only one answer, and more and more astronomers are inclined to accept it. The heart of the Galaxy, in the centre of Sagittarius A West, may contain a Black Hole.

Now that we have made up our minds, let us move on, making our way through the compact cluster of stars. Ahead of us there is an incredible scene. Matter is swirling around at tremendous velocity, forming what is termed an 'accretion disk' at a very high temperature, which allows it to emit both radio waves and very short waves. We seem to be watching something in the nature of a cosmic plug-hole, and because it is spinning rapidly the shock-waves around it are able to drive streams of gas outwards. The swirling gas and dust are making their final bid to avoid being dragged over the Black Hole's event horizon into the mysterious region from which there can be no return. Some of it may escape, to produce the expanding arms and clouds through which we have passed; much of it will not, and as soon as it enters the Black Hole it is lost to the outer universe for ever.

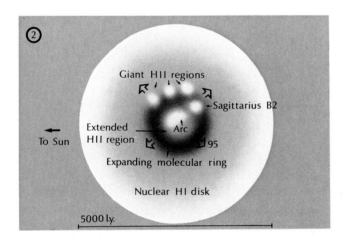

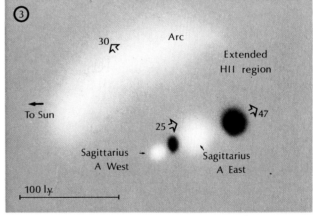

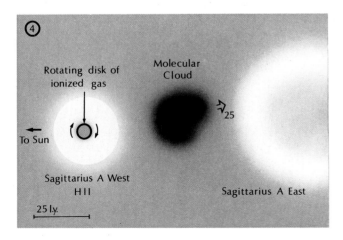

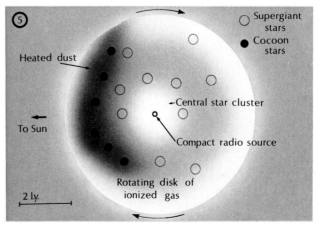

A Globular Cluster. We see this remarkable system from a vantage point on a hypothetical planet; there are globular clusters around the edge of the Galaxy, but, so far as we know, none near the centre of the system.

Now we can come back to the problem of the missing mass. If the Black Hole has about 5,000,000 times the mass of the Sun, the calculations fit, and we can see why the star cluster surrounding the compact central core is not dispersing. In fact, there seems no alternative to supposing that the core does contain this amount of mass, and nothing but a Black Hole seems to be adequate. The more material it draws in, as is happening every moment, the more massive the Black Hole becomes, and the larger it grows. It has even been said that a Black Hole is a cosmic cannibal with an insatiable appetite. It can take in not only thinly spread material, but even complete stars. And when it has collected enough new mass to become particularly active, as might happen every few million years, it could cause explosions violent enough to send arms and clouds speeding outwards as the molecular cloud between Sagittarius A East and West is doing now.

There is no point in our going further, even on our light-beam. We can penetrate as far as the swirling material; beyond, we can see nothing, any more than we could do when closing in upon the much less massive Black Hole of Cygnus X-1. We have travelled as far as we dare.

If our Galaxy has a central Black Hole, then presumably the same may be true of other galaxies, and even the much less massive systems that we call globular clusters, next on our list of stopping-points. Presumably, then, Black Holes at galactic centres were formed fairly early in the evolution of the galaxies themselves, and this is logical enough, since it is at the centre that the pressures and temperatures would be at their highest. As time passes, the central Black Holes will grow and grow, swallowing up more and more material; one or two alarmists have suggested that given sufficient time, the Black Hole in Sagittarius A West may become so greedy that it will pull in all the stars in the Galaxy, including our Sun. Nothing of the Galaxy we know would be left.

Frankly, this seems to be taking speculation much too far. Sagittarius A West is over 30,000 light-years away from us, and the time taken for it to expand sufficiently to engulf the Sun would have to be reckoned in millions of millions of years, even if it were possible at all – and the Sun will not survive for as long as that in any case. So we need have no qualms on that score, and the chances of our Solar System being approached by a separate, single Black Hole wanderer appear to be extremely remote.

We have reached the edge of the galactic core; we are only a few hundreds of millions of miles from it, and this is our limit, even on our light-beam. To continue our flight we must retrace our steps. Leaving the maëlstrom of the Black Hole, we pass once more through the inner star cluster with its crowded suns; next we bypass the expanding molecular cloud and the old supernova, Sagittarius A East; we manoeuvre our way through the Arc, then through the H.II regions and their clouds before emerging from the core region by way of the normal-hydrogen disk. We leave the expanding exterior arms behind, and pause to look back at the fascinating heart of the Galaxy, now hidden again. Eventually we return to regions from which only the glorious star clouds of the Archer indicate the direction in which the core lies. We are now on our way not to the middle of the Galaxy, but to its edge. Only then can we leave the Milky Way, and start the next part of our journey.

9 Gas Clouds and Spirals

We have left the centre of the Galaxy; now it is time to make for its edge, and see what we can find there. We will also start to see the Galaxy as a whole, which is something that Earth-bound astronomers can never hope to do. In particular we will encounter the globular clusters, great spherical-shaped systems which seem almost like miniature galaxies.

Actually, not all the globular clusters are so remote. The nearest of them is known as NGC 6397 (the letters standing for the New General Catalogue, drawn up nearly a century ago by the Danish astronomer J. E. E. Dreyer). It is not one of the larger globulars, and it is not visible from England, because it lies in the southern constellation of Ara, the Altar, and never rises above our horizon. As we pass by we can see that it is highly concentrated towards its centre, and there seems to be a central 'clump', which is decidedly bluish in colour. Astronomers believe that NGC 6397 is very ancient. Its age has been estimated as 17,000 million years, but this raises new problems, because the universe itself may not be as old as that. However, there are bigger and better globulars to come, so we will not call in on NGC 6397; we will continue on our way towards the edge of the Galaxy.

One thing we do notice is that the main Galaxy is flattened, with a central bulge – the mysterious core, which we have just left. Looking along the main plane of the Galaxy

NGC 4603. Far away we can see other galaxies, some of which are spirals – and in which presumably, star formation is going on in just the same way as in our own Galaxy.

means that we see many stars in almost the same direction, and this causes the Milky Way band which we see from Earth. Of course there are stars both above and below the main galactic plane, but there are fewer of them, and this adds force to the suggestion that during its early stages the Galaxy collapsed from an irregular cloud of material into a disk-shaped one.

This can happen, too, on a smaller scale. At about 10,000 light-years from Earth we come to an object known as MWC 349, which is decidedly unusual; it is made up of a very luminous star surrounded by a disk of glowing gas. The star itself is about 90,000,000 miles in diameter and 30 times as massive as the Sun; the disk is some twenty times the size of the star, and at its outer edge the thickness is about the same as the star's diameter. Evidently we are looking at a true cosmic infant, perhaps only about 1,000 years old. It can hardly go on shining for more than 100,000,000 years, because it is using its fuel at so furious a rate. Quite possibly the luminous disk is the inner part of a larger surrounding disk of dark material in which planets have already been born. The shining section is wedge shaped, and joins the brilliant surface of the star.

If this is correct, then with MWC 349 we have found a planetary system that is just forming. It is not the same as the Solar System, because the parent star is so much more massive and energetic; we can hardly believe that life will have time to evolve on any of its planets before the star explodes, but we cannot be sure.

There is more thinly spread material in the main plane of the Galaxy than elsewhere, so let us leave this region and continue our journey well away from the mid-plane. Here there are fewer solid particles ('grains'), but there are a great many atoms and molecules, and we can see that even here the space between the stars is not empty. True, the material is very tenuous even in the clouds of cold hydrogen which emit the 21-centimetre radio waves, but it exists, and it is highly significant. If we are carrying analytical equipment (much more sensitive than anything we could build on Earth as yet, of course), we will be able to identify complex molecules, including organic ones. Alcohol, for instance, is present in space, and it has been calculated that one single cloud could be used to make more whisky than Earthmen have consumed throughout their entire history (though suggestions that there must, somewhere, be a cosmic bar are not to be taken seriously!).

Impression of our Galaxy – from beyond. We are leaving our Galaxy behind; looking back, we will see it for the first time in its true form – a well-marked spiral.

Let us consider these organic molecules. Living things are made up of this material. This means that we cannot definitely rule out the possibility that life is created in space, and this is precisely what has been suggested very recently by two eminent astronomers, Sir Fred Hoyle and Dr Chandra Wickramasinghe. They point out, quite logically, that the production of living from non-living material involves a whole series of coincidences, each of which is in itself very unlikely, and they do not believe that the whole series could have been carried out on a small planet such as the Earth. They maintain that it needs the whole resources of space. In this case, life on Earth did not begin here; it was brought to our world from beyond the Solar System (Hoyle and Wickramasinghe regard a comet as an ideal carrier), and took root here simply because the Earth was suited to it. Other worlds would be similarly 'seeded', but not all would be welcoming. The Moon would not be suitable; its lack of atmosphere alone would mean that any life transported from space would be unable to survive there.

The theory has important implications. Space is so large that if it is correct, then every planet must have been 'seeded' at one time or another, and this means that every planet capable of supporting life will in fact do so. Few astronomers agree with Hoyle, and most are openly sceptical. But it is at least a possibility, and the existence of organic molecules in space cannot be questioned, even though anything of the sort would have been ridiculed a few decades ago.

We have left the centre of the Galaxy in a direction which will take us to a particularly fine example of a globular cluster, M.13 in the constellation of Hercules. It is not the closest of the globulars – that distinction belongs to the system in Ara – and neither is it the brightest; there are two in the southern sky, Omega Centauri and 47 Tucanæ, which surpass it in size and brilliance, but M.13 is conveniently placed for observers in the Earth's northern hemisphere. With the naked eye it can just be seen as a dim patch; telescopes show that it is made up of stars, though near the middle of the cluster the stars are so packed that they merge into a general glow.

Moving onwards, we start to resolve even the centre, and we can see that the brightest stars in M.13 are orange or red. This means that they must be old, and there is not much interstellar matter, so that apparently star formation has come to an end there; all the star-producing material has been used up, so that the globular cluster itself must be ancient.

How many stars does it contain? Many thousands; perhaps more than 500,000. It is a huge system, and its diameter is of the order of 100 light-years. The stars in it must be relatively close together. We have come across this sort of situation before, in the cluster adjoining Sagittarius A West, but here we are dealing with a different kind of system, and there is no definite indication of any central Black Hole.

As we come up to the edge of the cluster, the aspect of the sky changes. Ahead of us there is a blaze of light; behind there are comparatively few stars, because we are nearing the boundary of the main Galaxy. Once inside M.13, the situation changes again. It would be difficult for an astronomer there to find out much about the region outside the cluster; the sky seems to be filled with stars, and the predominant colour is orange, though there are some hotter white and bluish stars as well.

Now we are nearing the centre, and the sky is more glorious still. There could be no true night on any planet orbiting a star in M.13, and instead of points of light we would see the disks of near-by stars, some of them within a few light-days. A few of these stars are variable. Watching carefully, we find

Messier 13 in Hercules; a magnificent globular cluster, with a total population of perhaps a million stars. A planet moving round a star near its centre would have no darkness at all.

that some of these are regular in their behaviour; they are Cepheids, and it was by observing them that astronomers on Earth were first able to measure the distance of the cluster itself. True, M.13 contains far fewer Cepheids than most globulars, but there are enough of them to be extremely useful.

From M.13, the Sun will be reduced to a very dim speck, so faint that it would be almost beyond the range of any instruments so far built on Earth. If we use our Super Telescope, we will be looking back 25,000 years to see an Earth upon which the great

ice-sheets extend well towards the Equator; men are purely nomadic, feeding upon animals such as mammoths and mastodons. There is no separate Britain; not until much later, at the end of the last Ice Age, did it separate from the mainland of Europe, when the ice melted and the sea covered the low-lying plain we now call the North Sea.

From Earth, the globular clusters are more numerous in the southern sky than in the northern. It was this which led the American astronomer Harlow Shapley, over 60 years ago now, to make two important discoveries. First, the Sun is not near the centre of the Galaxy, as had been previously thought. Second, the globular clusters form a kind of outer framework to the main system, and most of them lie round its edge. Using these facts, Shapley was then able to give a reasonably good estimate of the diameter of the Galaxy itself. We now know that the true value is around 100,000 light-years.

On our light-beam, we can change our direction so as to make a circuit of the Galaxy. It will take us a long time, but we will find more than 100 globulars, with scattered stars, which make up what is called the galactic halo. All are moving round the centre of the Galaxy, but the halo objects travel in steeply inclined paths instead of keeping to the region of the main plane.

The Pleiades or Seven Sisters; a young cluster, and as we draw inwards we can clearly see the reflection nebula – which is gas and dust spread between stars which are not hot enough to make it shine on its own account, as happens in Orion's Sword.

Now let us move further out, to the edge of the halo, and see what a bird's-eye view of the Galaxy will tell us.

As expected, the general shape is spiral, with rather loose arms. The resemblance to a Catherine-wheel is unmistakable; the rotation is slow, as is inevitable in so large a system, but away from the centre the traffic rules are the same as those in the Solar System, with the more distant objects moving more slowly than those which are closer to the core. Let us pick out the Sun, one of many stars just outside the edge of a spiral arm. It is travelling round the centre at a speed of about 1,300 miles per second, but even so it will take about 225,000,000 years to complete one orbit. This period is sometimes called the 'cosmic year'. Assuming that the Sun is 5,000 million years old (a reasonable estimate), it has therefore had

time to complete more than twenty circuits. One cosmic year ago, the Earth was in the geological Carboniferous period, when the coal forests were being laid down; there were swamps, marshes, ferns, huge horse-tails, and insects such as dragonflies, while the most advanced life-forms were am-phibians, and even the dinosaurs lay in the future. Two cosmic years ago life was in its primitive stage, with only simple, tiny creatures in the oceans. It is interesting to speculate as to what conditions may be like one cosmic year hence. . . .

If we use our time-travelling device (in the best Dr Who tradition) and speed every-thing up, we will see that as the Sun moves round the galactic centre, it catches up stars which are further away from the core and is itself overtaken by stars which are closer in. This effect was detected long ago, and was one of the ways in which astronomers were able to work out the overall shape as well as the total mass of the Galaxy, even though there are so many complications that the problem is immensely difficult.

The spiral arms present problems, too. Unquestionably they are 'trailing' as the Galaxy rotates, and it would seem that they ought to wind up and disappear in a com-paratively short time – certainly less than 5,000 million years, and we cannot believe that the Galaxy is less than twice this age. In fact, spiral arms should not exist. Yet they do; and we have to find some way of ex-plaining not only why they persist, but how they came to be formed in the first place.

There may be a clue to be found from the globular clusters, so let us look quickly at another of them: Omega Centauri, in the southern sky, which is larger and richer than M.13 and also contains many more variable stars. It is not rotating, and neither has it much interstellar material left, in

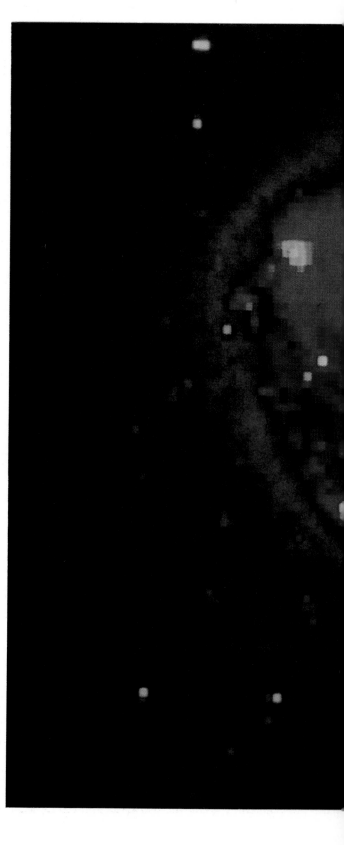

The galaxy NGC 5364, about 75,000,000 light-years away in the constellation of Virgo. Like our Galaxy it is a spiral but its arms are more tightly wound, its nucleus correspondingly larger.

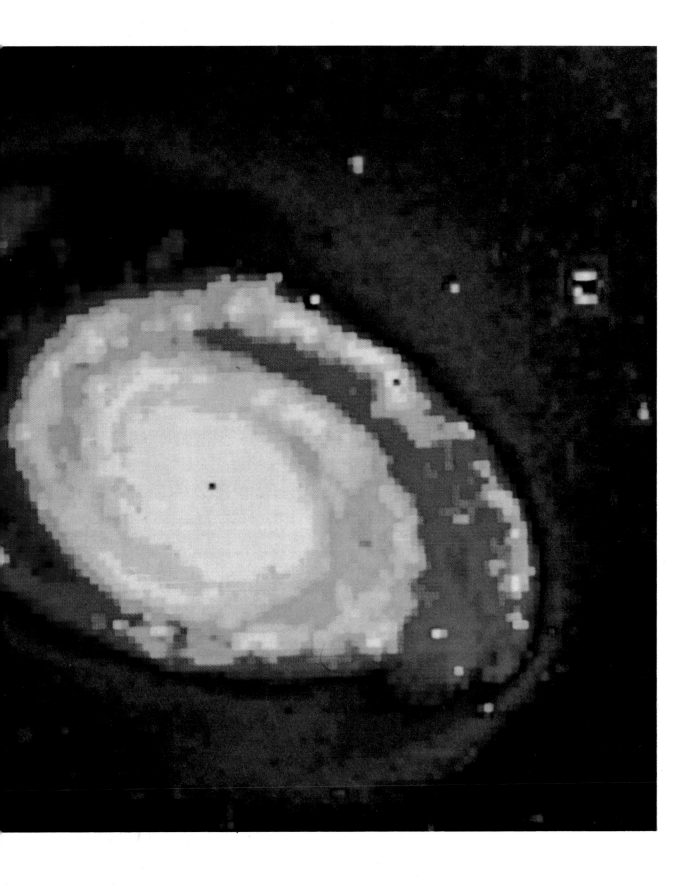

Dome of UKIRT, the United Kingdom Infra-Red Telescope, on Mauna Kea in Hawaii; the altitude is almost 14,000 feet, and it is possible to study infra-red radiations from space which at lower altitudes are absorbed by water-vapour in our own air.

which it is like all other globulars. Without rotation there can be no spiral arms, and Omega Centauri and its kind have become spherical. Later on we will come to whole galaxies that similarly lack rotation – and that have no spiral arms.

Go back in time to something like 10,000 million years (a few million years more or less makes no difference to our argument). The Galaxy was young. It consisted largely of hydrogen and helium, and it was collapsing; eventually it became disk-shaped, and during this process it started to rotate. Just how this happened is unclear, but it does

seem that any chance irregularities will start up a rotational movement, and once rotation has begun there is nothing to stop it.

At about the same time, what may be termed density waves were set up in the collapsing, rotating Galaxy. Inside these density waves the material was less tenuous than elsewhere, so that first gaseous nebulæ and then stars began to form. Some of these new stars were very massive and luminous, so that they used up their nuclear fuel relatively quickly and then exploded as supernovæ, hurling much of their material away into space. As we have found, heavy elements are built up inside massive stars as they evolve, and so when the supernovæ exploded the interstellar medium was enriched by these heavy elements, though hydrogen still remained more plentiful than anything else. Fresh stars began to build up from the ejected material, but this time they were less rich in hydrogen and richer in heavy elements. They were in fact second-generation stars.

Not all the stars originally produced were large and highly luminous. Smaller, less massive stars also condensed, and did not explode as supernovæ; instead they ran through their evolution in a much more leisurely and sedate manner, ending up as white dwarfs. Meanwhile, the density waves moved on, rotating round the galactic centre more slowly than the actual stars. Consequently, the most favourable locations for star formation shifted along together with the density waves.

Now things start to become more logical. A spiral arm will cease to be evident once its highly luminous stars have disappeared (either as supernovæ or as Black Holes), and new spiral arms will be produced by the slower-moving density waves. If this picture is correct – and most astronomers now believe so – then no individual spiral arm is a permanent feature. As old arms vanish, new ones appear to replace them. If we could project ourselves thousands of millions of years into the future, we would find that the Galaxy is still spiral, but the arms will not be the same as those we see today.

Our Sun is not a first-generation star. It is not old enough for that. So far as we can judge, the age of the Galaxy itself is at least twice as great. But there seems little doubt that the Sun began its career inside a density wave or spiral arm, and that it was produced from the material in a gaseous nebula. At present it is not inside a spiral arm, though in the future it may well overtake a new density wave, and will again be in a region of the Galaxy in which conditions for star formation are good.

We are almost ready to leave the Galaxy; but before doing so, we must make one last

The UKIRT, itself a telescope with a 150-inch mirror, built for infra-red studies. It is the largest of its kind and has proved to be so good that it can undertake ordinary work as well.

visit, even though it means retracing our steps. We must not omit to look at the Great Nebula in Orion, because it is the most famous example of a stellar nursery known to us. It is no more than 1,500 light-years away, and it is clearly visible with the naked eye below the three bright stars that make up the Belt of Orion the Hunter.

The nebula itself, Messier 42, is only the brightest part of a huge cloud which covers most of the constellation of Orion. It shines because of the bright stars mixed in it. The most obvious of these make up the multiple system of Theta Orionis, known as the

The Horse's Head Nebula. We can well see why this nebula has its name; it does resemble the head of a knight in chess. The very powerful star is Alnitak or Theta Orionis, at least 30,000 times more luminous than the Sun.

Trapezium because of the arrangement of its four main components. All the Trapezium stars are hot and bluish white, so that they not only illuminate the nebular material in their neighbourhood but also make the material emit a certain amount of light on its own account. If we travel in towards the Trapezium, we will find that opaque though it seems, the nebula itself is almost incredibly rarefied. If we could take an inch core sample right through it, the total weight of material collected would weigh less than a pencil.

Strange things are happening inside the nebula M.42. We can be confident that star formation is going on, because we have actually seen the appearance of fresh stars which have blown away their dusty cocoons and have started to shine. There are also young, unstable stars which fluctuate irregularly. Unfortunately, we cannot see deep inside the nebula, and this is tantalizing, because we know that there are mysterious objects inside that would tell us a great deal if only we could catch a glimpse of them. One such object, in particular, would be highly informative.

If we cannot see it, then how do we know that it exists? The answer is that it sends out infra-red radiation, which, unlike visible light, is not blocked by the dust in the nebula. Infra-red can be collected in much the same way as visible light, and by now there are several major telescopes devoted to this kind of work; pride of place must go to UKIRT, the United Kingdom Infra-Red Telescope, which has been sited on the top of Mauna Kea in Hawaii and which has a main mirror 150 inches in diameter. (Mauna Kea, at an altitude of almost 14,000 feet, is ideal for the purpose; at lower levels, infra-red rays coming from space are blocked by the water vapour in our own air, but at 14,000 feet most of the atmospheric water vapour lies below.) The UKIRT cannot produce a visible picture in infra-red – though it is also used for ordinary visual work – but it can detect objects that would otherwise be beyond our range. The most interesting object inside M.42 is known as the Becklin-Neugebauer Object, after its co-discoverers, though most people refer to it simply as BN.

In our all-powerful spaceship, we plunge past the Trapezium and enter the great nebula itself. The Trapezium stars are infants, perhaps only 100,000 years old, but they have had time to 'burn a hole' in the dark cloud, blowing away the obscuring dust and rendering themselves visible; BN is much deeper in the cloud, and is permanently hidden. As we move inwards, the cloud becomes not only dark but also cold, at a temperature of well below −300 degrees Fahrenheit. Our infra-red equipment tells us that BN is ahead; so, too, is another infra-red source, the Kleinmann-Low Object or KL, which is fairly close to BN but is not identical to it.

There are two theories about BN. Either it is a very young star, still condensing and not yet hot enough to emit visible light (though strong in infra-red), or else it is a very large, powerful star which would be brilliant in our skies were it not immersed in nebulosity.

If BN is a super-luminous star (as many astronomers, though not all, tend to believe), then we are faced with an extraordinary situation. Here, deep inside the darkness of the cloud, is one of Nature's searchlights, doing its best to penetrate the cosmical fog and make its presence known in the universe beyond. But it is doomed to failure. It will not last for long, because it is using so much of its nuclear fuel, and long before it can burn a hole through to outer space it will have died, either by exploding as a supernova or else collapsing into a Black Hole. We on Earth can never hope to see BN.

KL is different. As we leave BN and move across to it, travelling through the gloom of the cloud, we come across material which is speeding along at up to 17,000 miles per hour, but of KL itself there is no direct sign. If it has just started to shine, then it will be blowing out material in all directions in a kind of 'stellar wind'; where the flow speed reaches the speed of sound, a shock-front is produced, and this could be the case with KL. Moreover, it now seems that there are several centres from which material is expanding quickly, so that perhaps KL is not a single star just being formed, but is made up of a whole cluster of proto-stars. This, too, is logical enough; we have excellent evidence that stars do form in groups. The open cluster of the Pleiades or Seven Sisters, prominent in the night sky throughout autumn and winter, is a particularly good example.

We can find out no more. BN and KL, those enigmatical though non-identical twins, guard their secrets well. So we will take our leave of them, passing out through the Orion cloud with its young, unstable stars and its brilliant Trapezium. As soon as we are clear, we see the stars of the main Galaxy again; our journey takes us past the main plane, the globular clusters and the halo stars, and out into the reaches of intergalactic space.

10 Island Universes

In preparing for our departure from the Galaxy, we have to make some mental readjustments. First, we will in future be dealing with distances not of a few tens or hundreds of light-years, but of thousands, millions and even thousands of millions of light-years. Second, our Super Telescope will no longer show the Earth we know, because as we travel outwards we will pass objects so remote that their light would take over 5,000 million years to reach us – and the Earth is not 5,000 million years old.

As we accelerate away from the Galaxy, something very unexpected comes into view. It is a globular cluster, but a globular with a difference. It is visible from Earth, and is No. 5694 in the New General Catalogue; it is exceptional because it is moving in a way which will lead to its escaping from the Galaxy altogether, to become what may be termed an 'intergalactic tramp'. Probably it has been pulled out of its original orbit by another, more massive globular; at any rate, it is on its way, and it will not return. From it, the sky would be peculiar. Near the centre there will be the usual blaze of light from the orange and red giant stars which are the globular's leading members; from a planet orbiting a star near the edge of the cluster, part of the sky will be swarming with brilliant stars while the other part will be almost blank. Since the Tramp contains little interstellar dust or gas, no new stars can be forming inside it, and eventually there will be nothing left but a dead, isolated system made up of white dwarfs and, presumably, Black Holes.

Tramps are not easy to find unless they are still in the neighbourhood of the Galaxy, because they are not powerful enough to be seen over millions of light-years. The first major galaxies which we reach are less than 200,000 light-years away, and both are so bright that they are easy naked-eye objects, though both are too far south to be seen from the British Isles or anywhere in Europe. We call them the Magellanic Clouds, since they were described by the great explorer Magellan during his voyage round the world in 1520, but they must have been noticed well before that. There are two Clouds, and we come first to the larger of the pair, which is a fair-sized galaxy in its own right; it is 30,000 light-years across. No spiral shape can be made out, and the shape seems to be to all intents and purposes irregular.

Why no spiral arms? The probable answer is that the Cloud is not rotating, so that there are no density waves in it. Yet star formation is certainly going on, and as we come closer we can see many features of familiar type. There are globular clusters, open clusters, gaseous nebulæ, planetaries, and stars of all kinds, from highly luminous supergiants down to cool dwarfs. There is even one known pulsar, detected in 1982 by astronomers in Australia. One star in particular in the large Cloud catches our attention: S Doradûs.

Impression of Centaurus A, the nearest large radio galaxy, once wrongly thought to be made up of two galaxies in collision. We can see the remarkable complexity of the system, which is much the finest example of this type of galaxy known to us.

Messier 32. As we pass by M.31, the Andromeda Spiral, we notice that it has two satellite systems, M.32 and NGC 205. M.32 is elliptical, and has no spiral arms, so that its leading stars are old and red.

It is yellowish, and immensely powerful; its luminosity is at least 1,000,000 times that of the Sun, and may be more. Apart from Eta Carinæ, it is the most powerful star ever found, and this means that it cannot shine for long in the way that it is doing now. A million years hence, and S Doradûs will certainly have collapsed, and in all probability nothing will remain of it but a Black Hole. From Earth, it cannot be seen without optical aid – a reminder of how distant the Cloud is.

Now we can make out another interesting feature of the Large Cloud: a nebula beside which M.42 in Orion would seem puny. It is known as the Tarantula Nebula because of its intricate, twisted shape; it is, of course, an H.II region, and the giant stars inside it are able to illuminate an area extending over 900 light-years. If the Tarantula were as close to the Earth as M.42, it would cast shadows, and the total mass of the hydrogen alone is equal to that of 500,000 Suns. The Small Magellanic Cloud, slightly further away, is of the same basic type, though without anything so spectacular as the Tarantula. There are plenty of Cepheids, which, as we have seen, were used by Miss Leavitt in her pioneer researches into stellar luminosities. The Clouds are engulfed in a vast cloud of rarefied gas, and both may well be moving round our Galaxy in the manner of satellites.

Beyond the Clouds we come to a region which is sparsely populated. There are a few very small galaxies on view, containing only a few million stars each and hardly more imposing than globular clusters; one of them – in the constellation of the Little Bear, as seen from Earth – is apparently almost on the verge of disruption. We are not sure how it is moving, but if it is incautious enough to come too close to the Galaxy it will cease to exist as a separate system, and its stars will be captured. This cannot happen for a very long time in the future, but it is not impossible.

As we continue to move outwards, our Galaxy shrinks in the distance, and within 1,000,000 light-years it is outshone by another system, M.31, the Andromeda Spiral, so large that even at its distance of 2,200,000 light-years from the Earth it is just visible with the naked eye. From Earth we see it as an oval. Travelling on our light-beam we can move away from its main plane so as to give ourselves a bird's-eye view, and M.31 proves to be another Catherine-wheel, with well-marked spiral arms and all the familiar types of objects, including the Cepheid variables which gave the first clue to its distance. Before these variables were

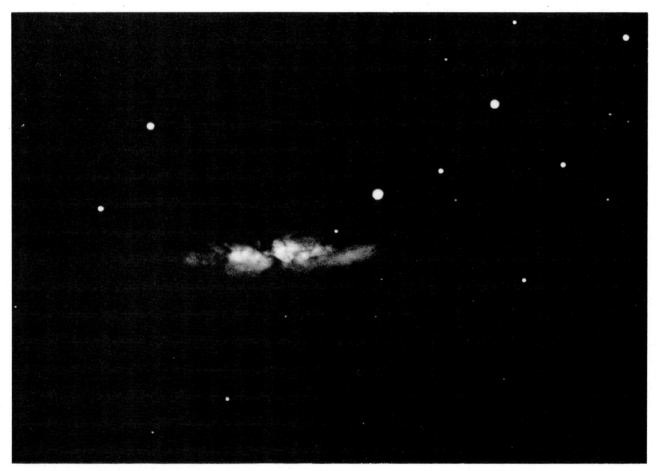

An exploding galaxy: M.82 in the Great Bear. We can see that tremendous disturbances have taken place inside the system, and that great streamers of hydrogen are still moving at very high velocities. There is no hint of a spiral form.

studied, M.31 was believed to be part of our Galaxy, but in 1923 Edwin Hubble used the great Mount Wilson 100-inch reflector to determine the distances of its Cepheids, and showed that it is in fact what may be called an 'island universe'. Its diameter is one and a half times that of our Galaxy, and it contains more than our Galaxy's quota of 100,000 million stars. In its halo, there are over 300 globular clusters.

Once inside M.31, we find that conditions are much the same as in our Galaxy. Stars of all kinds, gaseous nebulæ, planetaries – all are present; but when we look for our Galaxy we are forced to use optical aid, because it is beyond naked-eye range. With our Super Telescope we look back at the Earth of more than 2,000,000 years ago. The last Ice Age has not begun; the world is warm, but of civilization there is no trace.

Radio waves have been picked up from M.31, and there is a well-defined nucleus which may contain a Black Hole, though it is impossible to be sure. And like our Galaxy, M.31 has two companion systems, one of which is irregular and the other elliptical, lacking spiral arms or star-forming material.

Away from the Andromeda system we come to another spiral, the Triangulum Spiral or M.33, which is smaller and fainter, with a mere 10,000 million stars. Its arms are looser and less well marked than those of our Galaxy or M.31, but they are quite recognizable. No doubt some of its stars are centres of planetary systems. If so, and we decide to visit one, we will find that the sky is dominated partly by the stars of M.33 itself and partly by the Andromeda Spiral, which is brilliant and imposing over a distance of only a few hundred thousand light-years.

Something else is starting to emerge. Dwarf galaxies are becoming scarcer, and we are conscious that we are about to leave a whole collection of systems of which our Galaxy, M.31 and M.33 are the senior members. They, together with more than two dozen dwarf galaxies, make up what we call the Local Group. It extends well out beyond the three spirals, and there is another galaxy, known as Maffei 1 in honour of its discoverer, which is a distinct puzzle. It is not a spiral; it is elliptical, and so presumably not a quick spinner, but it is not the sort of system which we would expect to find so close to us. The trouble is that from Earth it is difficult to see at all, because it is hidden by the dust in the main plane of our Galaxy. If its distance is only 4,000,000 light-years, we must regard it as a member of the Local Group, but there is a great deal of uncertainty about both it and another galaxy, the spiral Maffei 2, which is further away but lies beside Maffei 1 in the constellation of Cassiopeia. Whether or not they are true members of the Local Group remains to be decided.

Beyond the Local Group there is an immense gulf, filled by nothing more than the incredibly tenuous material which seems to pervade all space. As we draw further and further away, the sky becomes almost starless; there may be the occasional intergalactic tramp in the form of a globular cluster, but little else. Eventually other galaxies begin to come into view, but there is a change in the overall situation. The Local

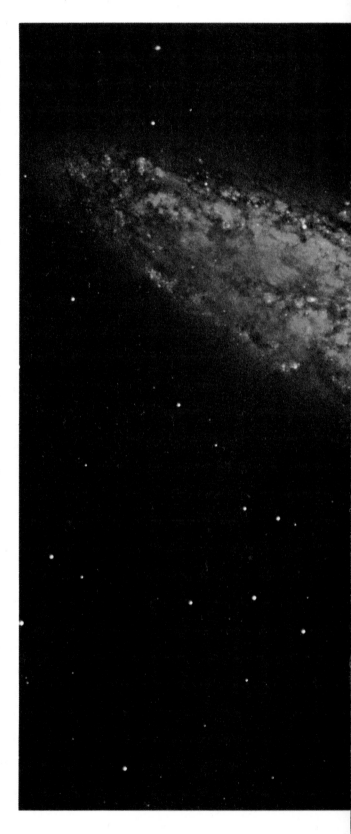

Spiral galaxy NGC 253, in Sculptor. From this angle, the spiral form is largely concealed, but seen from 'above' or 'below' the spiral arms would be very prominent.

Group is stable, so that its members keep together as a family. Beyond it, all the galaxies, whether large or small, spiral, elliptical or irregular, are rushing away from us. The entire universe is expanding.

This seems staggering. Can it be that the Local Group is particularly unpopular, so that the other galaxies and groups of galaxies are doing their best to run away from it? Not so. The expansion is, in every sense of the term, universal. Every group of galaxies is racing away from every other group, and the greater the distance, the greater the speed of recession.

We have known about this for some time. The pioneer work was carried out at the Lowell Observatory, Arizona, just before World War I, when E. C. Slipher used the powerful telescope there to examine the spectra of the galaxies. It was then that he detected what we now call the red shift. Its interpretation was delayed for some time, because at that stage it was not generally believed that the spirals were external systems or 'island universes', but when the interpretation was made, following the work of Edwin Hubble and his colleague Milton Humason in 1923, it caused a complete somersault in all our ideas about the universe.

As we continue our journey, let us make use of a spectroscope to split up the light of a spiral galaxy, M.81, which from Earth is seen in the constellation of the Great Bear. When the light is passed through a glass prism (or an equivalent device) it is spread out into a band of all the colours of the rainbow, from red at the long-wave end through orange, yellow, green and blue to violet at the short-wave end. A hot solid, liquid or high-pressure gas will always produce this sort of rainbow, known as a continuous spectrum, but a low-pressure gas will not; instead, it will yield bright, isolated lines of different colours. Each line is due to one particular element or group of elements, and is a distinctive trade-mark which cannot be duplicated by anything else. Thus, if we see two bright yellow lines in a certain definite position, we know that they must be due to sodium, one of the two elements making up common salt. As soon as these lines are seen, we know that sodium is present. Other elements produce other lines; iron, for instance, is responsible for thousands.

With a star, these two types of spectra are combined. The hot surface produces a rainbow. Surrounding the surface are layers of thinner gases. On their own, these gases would yield bright lines, but against the rainbow background the lines are reversed, and appear dark. This does not matter; they can still be identified, because their positions and their intensities are unaltered. Thus in the yellow part of the Sun's spectrum we see two dark lines which betray the presence of sodium. By now, over 70 different elements have been identified in the Sun.

The spectrum of a galaxy is made up of the combined spectra of millions of stars, and is bound to be something of a jumble, but the principal lines are clear enough, and can be identified. Predictably, we find lines due to the familiar elements – hydrogen, iron and all the rest; but there is another effect, too, which can tell us not only which substances are present, but also how the galaxies are moving. This is the Doppler Effect, first pointed out in 1842 by the Austrian physicist Christian Doppler.

Consider a police car or an ambulance, moving quickly along with its siren screeching. The note will be high pitched as the vehicle approaches, because more sound-waves per second enter your ear than would be the case if the vehicle were

Impression of a planet in orbit around S. Doradûs. The heat is so intense that the planet may never have cooled down. The only relief from the heat would be an eclipse by a moon of the planet, then the spectacular corona would fill half the sky.

motionless; the sound-waves are apparently shortened. Once the vehicle has passed by, and has started to move away, the note of the siren drops, because fewer sound-waves per second reach your ear, and the wavelength is lengthened. Much the same is true of light. With an approaching source the wavelength is shortened, and the light is 'too blue'; with a receding source, the increased wavelength makes the light 'too red'. This shows up in the positions of the lines in the spectra of stars or galaxies. If the lines are shifted over to the red end of the rainbow, the object is receding.

The Whirlpool. We can see this spiral galaxy face-on from Earth and can enjoy the full beauty of it. As we pass by on our light-beam we notice that it is connected to a smaller galaxy by a definite link or branch.

This is what Hubble and Humason showed with the galaxies. Apart from the members of the Local Group, every galaxy showed a red shift in its spectrum. The amount of the shift gave a clue as to the velocity, and it was found that the more remote galaxies showed the greater red shifts. The invariable rule was 'the further, the faster', and it was this which proved that the universe is in a state of expansion.

As we travel along on our light-beam we bypass other galaxies, but let us concentrate upon M.81 and its neighbour M.82, which as seen from Earth lie close together. Each is about 10,000,000 light-years away, and each is receding, but they are not alike. M.81 is a typical spiral, but M.82 is irregular. Closing in, we see that there is great activity going on inside M.82. There are not many hot blue or white stars, but there are huge streams of hydrogen gas, moving at speeds of up to 600 miles per second; evidently there has been a violent explosion in the centre of the system, and this is confirmed by the fact that M.82 is sending out a strong flux of radio waves. From the movements of the hydrogen structures we can calculate that the explosion happened about 1,500,000 years before the indications of it reached Earth and M.82 is 11,500,000 light-years away from us. Apparently the radio emissions are due to electrons which are spiralling round in a powerful magnetic field, a process known as synchrotron emission (also very marked in the Crab Nebula, incidentally). If this interpretation is correct, then M.82 is an exploding galaxy. It is true that not all astronomers agree, but at any rate M.82 is a remarkable system.

Next we must call in on M.51, the spiral galaxy which has been nicknamed the Whirlpool because of its beautiful form; it was the first spiral to be seen as such, when Lord Rosse observed it in 1845 with his newly completed telescope, at that time much the most powerful ever made. M.51 is

13,000,000 light-years from Earth, and a superb sight; but we must pass on to something quite different, Centaurus A, which is unlike any system we have yet encountered. It is 16,000,000 light-years from Earth, and gives the impression of being a mixture of two galaxies; one bright, the other dark and elongated, cutting across its companion and giving the impression of slicing it in half. Centaurus A is a radio emitter, exceptionally powerful in the long wavelengths. It also sends out X-rays, and now and then outbursts of X-radiation are detected, perhaps indicating that an entire star or even a cluster of stars has been swallowed up by a massive Black Hole at the centre. The total luminosity of the system is 100,000 million times that of the Sun, and it is racing away from us at nearly 300 miles per second.

It used to be thought that Centaurus A, and others of its kind, were colliding galaxies, passing through each other in the manner of two orderly crowds moving in opposite directions. The individual stars would seldom score direct hits – they are too widely spaced for that – but the intergalactic material would be colliding all the time, and this, it was thought, could explain the strong radio emission. Further studies showed that this could not be so. There was much too much radio emission, and Centaurus A is now regarded as a single galaxy, though an extraordinary one.

Collisions do occur occasionally. One such encounter has produced the remarkable Cartwheel Galaxy, which has a diameter of about 170,000 light-years. A few hundred million years ago it was invaded by a much smaller galaxy, which passed close to the centre of the larger system and can still be seen. But just as stars seldom collide, so the galaxies generally keep at respectful distances from each other.

Travelling onwards, we approach another group of galaxies, the Virgo Cluster, which is much more populous than our Local Group; instead of a few dozen galaxies at most, it contains hundreds. The distance has been given variously as between 50,000,000 and 65,000,000 light-years. If the latter estimate is correct, then our Super Telescope will show the Earth at a significant phase in its history. As we focus on to the landscape, we see that the most advanced life-forms are very different from those of today. It was 65,000,000 years ago that the dinosaurs died out, to be succeeded by the much smaller, much more agile and much more intelligent mammals, from some of which we ourselves are descended.

It would be intriguing to have the chance to look back at the Earth of this age. The Cretaceous period was ending; at its close, the dinosaurs disappeared – not gradually, but very suddenly on the cosmic time scale. Nobody is sure why it happened. Some authorities believe that the Earth was hit by an asteroid large enough to cause an abrupt change in climatic conditions, and that the dinosaurs were unable to cope. Others believe that the dinosaurs, which, after all, had been lords of the world for well over 100,000,000 years, had simply come to the end of their evolution. In any case, they vanished, and the stage was set for the gradual emergence of Man.

If a radio astronomer living on a planet in the Virgo Cluster had transmitted a message at the time when the dinosaurs were dying out, it would only now be reaching us. And if another message were transmitted now, we would not receive it until AD 65,000,000. It is a long time to wait.

The most important member of the Virgo Cluster is M.87, a giant system more or less spherical in shape. It is a particularly powerful source of radio waves, and also transmits in most other regions of the spectrum, down to the short X-rays. We can see a peculiar jet of material extending from it; is it made up of stars, or of dust and gas?

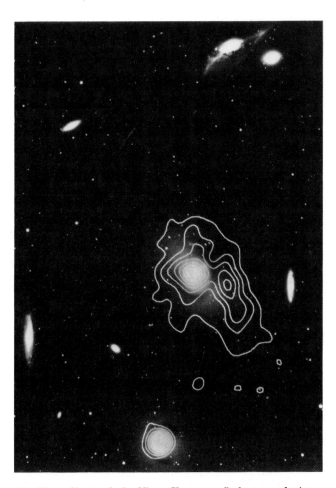

The Virgo Cluster. In the Virgo Cluster we find many galaxies, including M.84 and M.86; radio waves are emitted (white lines show the positions of the radio sources). The Virgo Cluster is very large, and is over 50,000,000 light-years away.

Almost certainly of both. We can see, too, that M.87 is truly a giant. Its diameter of 120,000 light-years is probably less than that of the Andromeda Spiral, but since it is spherical it is much more massive, and could match many thousands of millions of Suns. Surrounding it are at least 800 globular clusters. The brightest stars in M.87 are red, so that again we have found a system in which the most luminous members are old, and where no new stars are being formed.

There are strong indications of something very massive in the core of M.87. It can hardly be anything but a Black Hole, vastly larger and more massive than the modest feature in Sagittarius A West. So far as we can tell, nothing else can account for the energy of M.87.

What else? In the Virgo Cluster there are not only ellipticals and normal spirals but also the peculiar barred spirals, in which the spiral arms seem to extend from the ends of a luminous bar through the centre of the system. Every type of galaxy is represented, and it may even be that the Virgo Cluster is the centre of a more extensive 'super-cluster' which includes our Local Group.

By the time we have travelled onwards to a distance of 350,000,000 light-years, we can use our Super Telescope to look back at the Earth as it used to be in the Devonian or Old Red Sandstone period, long before the age of the coal forests. Plants had developed as tall tree-ferns; the sea contained fishes, but there were no advanced life-forms on land, and even the amphibians had not appeared. In time, the Devonian period is much more remote from the dinosaurs than the last dinosaurs are from us.

At 350,000,000 light-years we come to another group of galaxies, the Coma Cluster. Here, too, we find systems of all types, and though the Coma Cluster is not the equal of Virgo it is much larger than the Local Group. Further still, at 1,000 million light-years, we find Cygnus A, once believed to be the result of a cosmic collision but now believed to be a single huge radio galaxy, pouring out energy at a fantastic rate. To either side of the visible galaxy there are intense radio sources, each 300,000 light-years from the system itself; presumably these have been ejected in some titanic explosion. From Cygnus A, our Super Telescope would show the Pre-Cambrian Earth, with no life apart from primitive organisms in the seas. We would be unable to recognize any of the continents or oceans, since the whole aspect of the world has changed completely since that epoch.

The Coma Cluster. By now we are moving well away into the depths of the universe. The Coma Cluster contains galaxies of various kinds; looking back, we can now hardly make out our Local Group.

The red shift shows that Cygnus A is receding from us at 9,000 miles per second. Moving at the speed of light we have little difficulty in catching up with it, but now a new thought strikes us. If the rule of 'the further, the faster' holds good, we must eventually come to a distance at which a galaxy would be receding at the full 186,000 miles per second. If so, then we would be unable to see it, and we would have reached the boundary of the observable universe, though not necessarily of the universe itself. Galaxies are not powerful enough to be detectable out to this limit, but we have other objects to help us in our search: the quasars. It will be fitting to use these enigmatical quasars to bring our journey to its climax.

11 To the Outer Limits

We are 3,000 million light-years from home. Looking back we can see nothing of our Galaxy, or even the Local Group, without optical aid. Using our Super Telescope, we focus upon an Earth where life has gained a foothold, but has yet to develop; the continents, quite different in outline from those we know, are still barren, and even the seas contain nothing more than very primitive organisms.

Meanwhile, our attention is riveted upon something else. From Earth it is detectable, and has been catalogued as 3C-273 – that is to say, the 273rd entry in the third Cambridge catalogue of radio sources – but it is not a normal galaxy. It is the brightest of the objects which have become known as quasars.

Galaxies show a wide range in size, mass and luminosity. The Seyfert galaxies (named in honour of Carl Seyfert, who first drew attention to them in 1942) are of particular interest, because they have bright, condensed centres and only weak spiral arms; most of them are powerful at radio wavelengths, and once again we are at a loss to explain their energy unless we assume that the core of a Seyfert contains a Black Hole. But 3C-273 is different. It is small, and until we come within reasonably close range it looks like nothing more than a rather ill-defined blue star, with a jet extending from one side.

Impression of Stephan's Quartet. There is a double galaxy at the centre of this cluster and for a while, it was called Stephan's Quintet. One of the galaxies was then found to be receding faster than the others, and does not belong to the group.

Casting our minds back, we recall the interesting story of how 3C-273 was tracked down. In the early 1960s, the best radio telescopes still had very poor resolution; they could indicate the rough position of a radio source, but they were quite unable to pinpoint it with any accuracy, so that comparatively few sources had been identified with optical objects. Of course there were some, notably the Crab Nebula, but the situation was unclear. One source, in the constellation of Virgo (but not associated with the Virgo Cluster of galaxies) was particularly strong, and, by a lucky chance, it lay in a position in the sky in which it could sometimes be hidden or occulted by the Moon. By timing the exact moment of the occultation, when the radio signals were abruptly cut off, it was possible to find out exactly where the source lay. This was done by Australian radio astronomers in 1963, and the source was found to be so close to a blue, starlike point that there seemed little danger of a mistake.

The 'star' itself was bright enough to be seen through a fairly small telescope, and the next step was to examine its spectrum. All the available information was sent from Australia to Palomar, in California, site of what was then the world's largest telescope, the 200-inch Hale reflector. Spectra were obtained, but proved to be baffling. The spectral lines did not seem to correspond with anything at all, and for a while the mystery was complete.

Then Maarten Schmidt, at Palomar, realized the truth. The spectral lines were

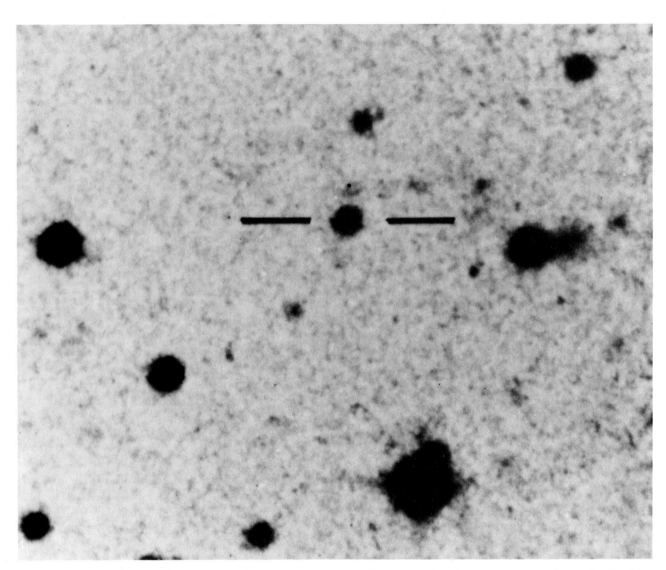

The ultimate limit – so far. The Quasar 2000-330, thought to be about 13,000 million light-years from Earth. With it, we come to the end of our journey on our light-beam; it is time to start coming home.

due to nothing more unusual than hydrogen, but they were so tremendously red-shifted that they had not been identified. The red shift indicated a distance of 3,000 million light-years. If 3C-273 were so remote, it would have to have at least ten times the luminosity of any known galaxy. It seemed impossible; but astronomers were becoming used to believing in impossibilities, and there seemed no doubt that the red shift really was the largest ever found. At first 3C-273 was referred to as a 'quasi-stellar radio object' or QSO, but after other similar objects were found the name was conveniently shortened to 'quasar'.

The velocity of recession was the most surprising feature. It amounted to one-sixth of the speed of light, or more than 30,000 miles per second, and when more and more quasars were discovered it became clear that 3C-273 was one of the closest members of the class. A few are nearer, but most are much more remote, and are receding at speeds which are appreciable fractions of that of light. Things were made easier by improvements in radio techniques. Not all quasars proved to be radio emitters. Many of them were radio quiet, but the more distant ones almost invariably were radio sources, and astronomers cast around for an explanation of how so much energy could pour out from an object which was no more than a few light-years in diameter at most.

The usual suggestions were made: chains of supernovæ, stellar collisions, collisions between galaxies, interaction between normal matter and anti-matter, and so on. None seemed to be particularly promising. It was then suggested that the red shifts in the

spectra of quasars were not Doppler Effects at all, but due to some other cause; possibly light could be sufficiently reddened during its escape from an intense gravitational field, or else reddened during its long journey to Earth. In America, Halton Arp of the Mount Wilson Observatory claimed that there were many cases of galaxies and quasars that were lined up and, therefore, presumably associated, even though they had completely different red shifts. If this were so, then quasars could still be comparatively near our Local Group, and though they would still be very strange they might not be so improbably luminous. Some astronomers, notably Sir Fred Hoyle, still hold this view, but on the majority opinion the red shifts are true Doppler Effects, and we have to explain the power of the quasars

Visually, galaxies seem to be associated with quasars. The galaxy (a) is an ill-formed spiral. The quasar (b) is arrowed in the photograph (B). The red shift of the quasar is greater compared with a spectrum produced in the laboratory (A).

A

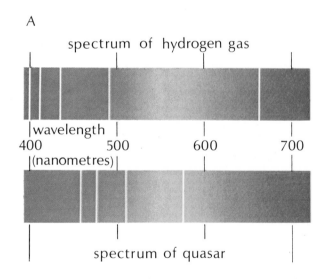

spectrum of hydrogen gas

| wavelength |
400   500   600   700
| (nanometres) |

spectrum of quasar

B

a

b

The 210-foot radio telescope at Parkes, Australia, used to discover the most remote object known – the Quasar 2000-330. The Parkes telescope is over twenty years old, but is still one of the best in the world.

somehow or other. Only now does it seem that we may be on the right track, so let us continue our journey and make a closer inspection of 3C-273.

Energy is being sent out at many different wavelengths. Visible light and radio radiation account for only part of the total; there are also short X-rays, indicating a very high temperature indeed, and at times there are surges of energy which show that violent disturbances are taking place. Yet there may be a fairly simple explanation. Quasars may be nothing more than the nuclei of exceptionally active galaxies, so that the fainter parts of the system are drowned by the brilliance of the core.

Even from close range we can see little sign of a galaxy associated with 3C-273, but presumably it must be there, and some other quasars do show tell-tale 'fuzziness'. One of these, MR 2251-178, is closer than 3C-273, at a mere 1,200 million light-years; the diameter of the associated galaxy is around 90,000 light-years, so that it is comparable with our Galaxy and much larger than the Triangulum Spiral.

Even so, we have still to explain the immense power, and once more we come back to the usual explanation – that in the core there is a super-massive Black Hole, swallowing up material and increasing its appetite as it does so. It is true that there is a modern tendency to regard Black Holes as 'remedies for everything', but unless we adopt the theory that quasars are fairly local there seems to be little alternative.

If we accept the majority view, it is evident that quasars give us our best chance of seeing into the remotest depths of the observable universe. Since they are so much more luminous than normal galaxies, they are detectable over a much wider

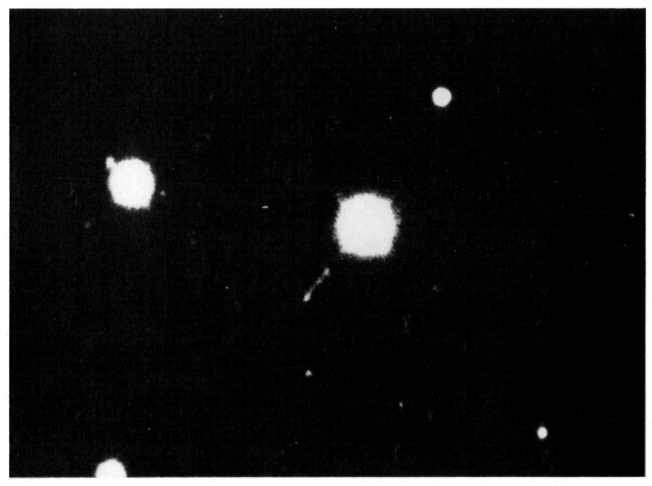

The Quasar 3C-273, brightest and first-discovered of the quasars. It shows a curious jet issuing from it, though even from close range we cannot make out precisely what it is.

range, and in our imaginary journey we can now move out to a distance at which no ordinary galaxy could be seen from Earth. We will make for the quasar OQ 172, discovered in 1973 by a team of astronomers at the Lick Observatory in America. According to its red shift, its distance must be well over 10,000 million light-years, and now, for the first time in our travels, we can use our Super Telescope to look back in the direction of the Solar System without the slightest hope of seeing anything at all. The Sun has not been born; neither has the Earth. The Solar System lies in the future. We have passed the Rubicon of time.

It may then be asked: 'What proof have we that OQ 172 really exists?' The answer is that we have no proof at all. We can say that OQ 172 existed 10,000 million years ago; but if it were suddenly snuffed out, we would not know for another 10,000 million years in the future. It is a sobering thought.

Also, there is the phenomenon of 'the further, the faster'. OQ 172 is receding at well over 160,000 miles per second, and if the general law holds good we are not so very far from the limit of the observable universe. But there, for over a decade, we stopped. Despite intensive searches, no quasar more remote than OQ 172 could be found, and this was puzzling. It seemed unlikely that OQ 172 really represented the limit; but although it was the most remote quasar known, it was by no means the faintest. Even if OQ 172 had been two or three times as remote as its red shift indicated, it could still have been detected.

Then, in 1982, a more distant quasar was at last found. Its radio signals were picked up at the Parkes Radio Astronomy Observatory in New South Wales, and the position of the source was sent over to the near-by Siding Spring Observatory, where there is one of the world's largest optical telescopes (and, arguably, the best). Almost at once the radio source was identified with a starlike point. It was catalogued as PKS 2000-330, and its red shift indicated a distance of at least 13,000 million light-years, with a velocity of recession of over 91 per cent of the speed of light.

PKS 2000-330 is our last port of call. There is every reason to believe that it, too, is the visible core of a more extended galaxy, with a super-massive Black Hole at its centre. But its significance is tremendous; we are at last coming to the outermost limits.

Remember, the expansion of the universe applies everywhere; every group of galaxies is racing away from every other group, and if we transfer ourselves to the neighbourhood of PKS 2000-330 we may be able to see across the boundary which restricts us from Earth. But again there is no proof, unless we are prepared to admit that the universe stretches outwards for ever.

Moreover, it is hopeless to pretend that we can be really precise when considering figures of this sort, and estimates of the distance of PKS 2000-330 range between 11,000 million light-years to more than 19,000 million light-years. If the lower value is correct, then it seems that the critical distance at which the recession of a quasar (or a galaxy) becomes equal to the velocity of light is approximately 15,000 million light-years. Further than that we cannot see, no matter how powerful and sensitive we make our equipment. Nature has set up an impenetrable if invisible barrier.

This, of course, assumes that the relationship between distance and speed of recession, known as Hubble's Law, holds good right out to the furthest limit. This may or may not be true; but if it is, then the observable universe has a definite boundary. It also has a definite age. Provided that Hubble's Law is the same now as it used to be when the universe was young, we can calculate backwards, so to speak, and deduce that the expansion did indeed begin 15,000 million years ago. This means that the universe is at least three times as old as the Solar System.

There is still much to be learned. Two decades ago, quasars were not only unknown but also unsuspected; they have issued a challenge, and though the challenge has been accepted we still cannot be sure that we are on the right track. Travelling through space can tell us no more; it is the moment to pack up our Super Telescope, our ultra-sensitive radio equipment and our X-ray detectors, and return home, which, even on our light-beam, will take us 13,000 million years. Our last journey will be not in space, but in time.

12 Journey Through Time

We have travelled from the Earth out to the remotest quasar yet discovered, PKS 2000-330. We have not begun our journey in a state of total ignorance, because although the bodies in the sky beyond the Solar System are out of our reach in our present state of technology, we can at least observe them and analyse their light. Time-travel is different. As I have said, going back into the past is probably one of the few things which we must regard as genuinely impossible. Going into the future is another matter, because of the time-dilation effect to which I referred before we began our journey; remember the twin paradox and the mu-mesons. On the other hand, there is no conceivable way in which we could emulate the scientist in H. G. Wells' famous novel *The Time Machine*, who simply enters a capsule and projects himself as far into the future as he chooses.

Therefore, when we decide to equip ourselves with a time capsule, we really are entering the realm of Dr Who, Captain Kirk and Lord Darth Vader. But there is no reason why we should not do so, always bearing in mind that we are describing something which cannot happen in reality. Therefore, we will board our time machine – the *Tardis*, if you like – and begin by going back to the start of the universe as we know it.

I say 'as we know it' because it is still not certain whether the universe had a definite beginning. According to the theory currently popular, it started between 13,000 million and 20,000 million years ago; from the evidence of the quasars and Hubble's

Law, we are entitled to take 15,000 million years as a reasonable compromise. Before that, nothing. Time and space did not exist. Then, abruptly, all the matter in the universe came into existence with a Big Bang. From our capsule we see a sudden blaze of light; the temperature soars to an incredible level of thousands of millions of degrees, and from the blaze we see particles together with photons (which are small 'parcels' of light) being shot outwards in all directions. The temperature falls quickly at first, then more slowly; all the time the universe becomes larger, and after only a few hours a certain amount of order begins to emerge from the chaos.

(This is all very well, but I am sure you will notice an inconsistency here. Where did the Big Bang happen? The official answer is that space came into existence at the same time as the matter, so that the Big Bang happened 'everywhere', and we would find it rather difficult to observe from outside, simply because there would be no outside! No matter; what is important is that the Big Bang started the phase of expansion which is still going on.)

Millions of years passed by. Chaos began to give way to order, until at last individual clouds of material began to form. The main substance was hydrogen, the simplest of all the elements; it is even possible that all material originally began as hydrogen, and was then built up into heavier elements. In

NGC 2359. One of the magnificent gas-and-dust nebulæ which we have encountered during our journey; it lies in a rich area.

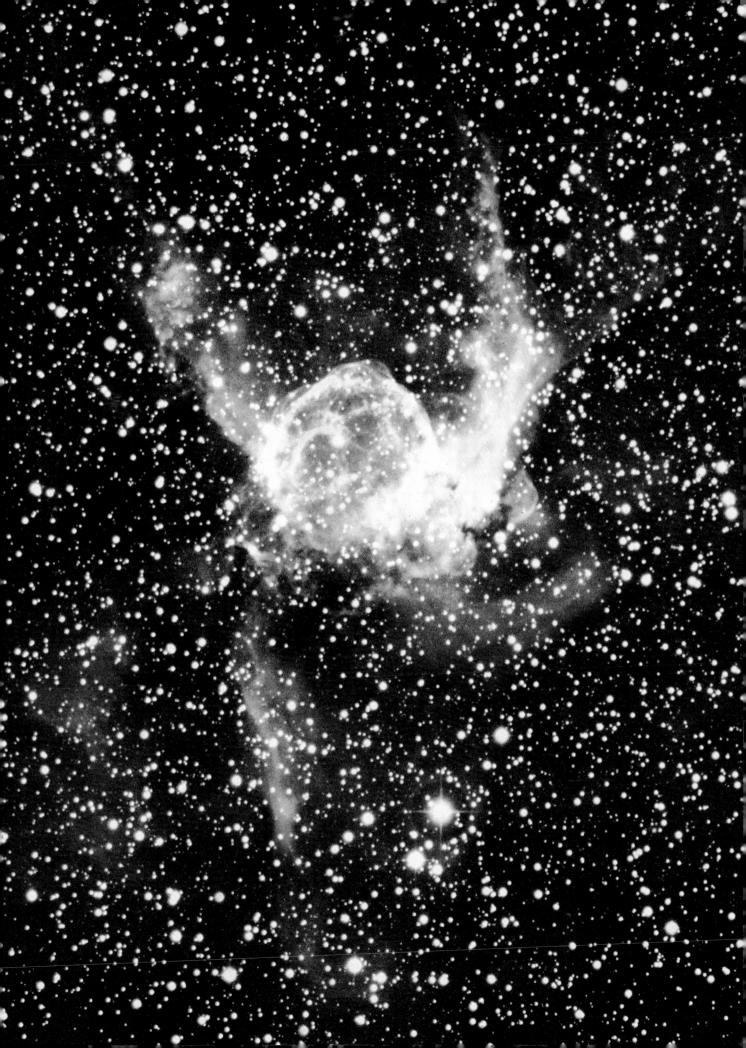

the course of time the embryo galaxies collected into groups, and star formation began. Some of the galaxies were rotating, so that they turned into spirals; others did not rotate – or, at least, very slowly – so that they produced the ellipticals, such as M.87 in the Virgo Cluster. Stars shone, some of them so luminous and massive that they lasted for only brief periods before exploding as supernovæ and hurling their material into space so that it could be collected into second-generation stars. Five thousand million years ago one star, our Sun, condensed out of nebular material, and shrank, together with the 'solar nebula', which became disk-shaped and then allowed planets to form. Away from the central part of the solar nebula the temperature was low; as we watch, we see large condensations forming, consisting mainly of the two lightest elements, hydrogen and helium, which

they can draw in as soon as their original cores have become sufficiently massive; these end up as the giant planets from Neptune to Jupiter. Closer to the young Sun the temperatures were much higher, and resulted in smaller, solid planets, with much of the hydrogen being driven out by the solar wind.

Slowly the Sun settled down on to the Main Sequence. It passed through the stage in which it was variable and unstable, and became sedate, though at first it was less luminous than it is today. By now all the planets already existed, and had begun to evolve in their different ways. The Earth was at its present distance from the Sun, but the Moon-Earth distance was less than it is now, and the Earth's day was shorter, probably only about twenty hours during the Cambrian period, which ended 500,000,000 years ago at the time when life had started in the warm oceans. Tidal effects drove the Moon outwards, and also slowed down the Earth's rotation. This slowing down still

Messier 87. A particularly powerful galaxy, as we see even when still a long way from it. It is a radio source, and is very massive; there is strong evidence of a Black Hole near its centre.

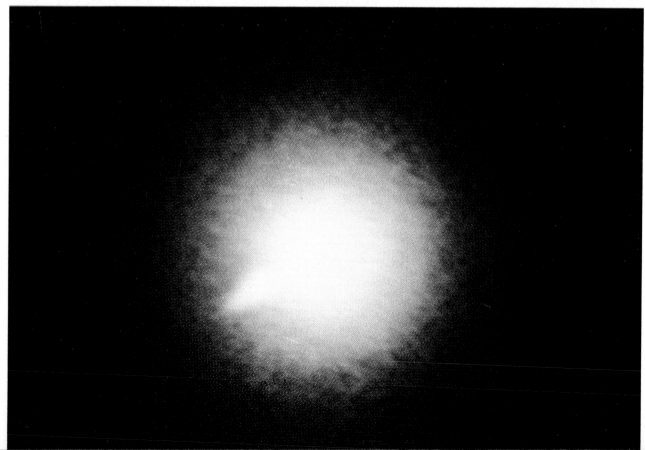

NGC 4755 Kappa Crucis, the Jewel Box in the Southern Cross.
We must not omit to visit it before leaving the Galaxy; it is so
named because it contains stars of contrasting colours. Note the
red giant star near the centre.

goes on, but it is very slight; only about one second in 50,000 years.

From our capsule we can follow the Earth through its life-story. The primitive single-celled organisms in the seas developed steadily; by the Silurian period, which began 430,000,000 years ago, fishes had appeared; then came the amphibians; during the Carboniferous period (345,000,000 to 280,000,000 years ago) the coal deposits were laid down, and then, between 280,000,000 and 65,000,000 years ago, came the long regime of the dinosaurs.

Actually most of them were vegetarian and harmless, though there were also flesh-eaters such as the tyrannosaurus, which must have been the most ferocious creature ever to have lived on Earth. At the end of the Cretaceous period the dinosaurs suddenly died out; mammals took over, until animals became recognizable, and, near the end of the story, Man appeared. Climates fluctuated; sometimes the Earth was cold, sometimes warm. The last Ice Age ended a mere 10,000 years ago, and in the future the climate may become chilly again. The cause of these glaciations has led to endless argument, though we may well believe that slight variations in the Sun's output have a great deal to do with them.

To put the time scale in perspective, it is rather useful to take the age of the Earth as being represented by one year. If so, we arrive at the following figures:

1 January:	Formation of the Earth
1 May:	Primitive marine organisms, like our modern single-celled algæ
25 October:	Many-celled life-forms in the seas
20 November:	Primitive fishes
30 November:	First amphibians
7 December:	Reptiles dominant
15 December:	The first mammals
31 December:	5 p.m. First hominids
31 December:	11 p.m. *Homo sapiens*
31 December:	11 hr 59 min 59.4 sec. Battle of Hastings

This brings us up to the present, and we can step out of our capsule to survey the world of AD 1983. But have we really solved the problem of the creation? The answer can only be 'No'. We may be able to prove that a Big Bang happened, but we have no idea why or how. In fact, nobody has ever discussed the *origin* of the universe from the purely scientific point of view. What we have been discussing is its *evolution*, which is by no means the same thing.

We do have evidence for the Big Bang, because it has been found that the universe is pervaded by weak radiation, indicating a temperature of three degrees above absolute zero (absolute zero being the coldest temperature possible), and this is almost certainly indicative of the remnant of the original Big Bang. But was this particular Big Bang the first of its kind?

Thirty years ago, a completely different picture was given by a group of astronomers at Cambridge University. To them, there was no Big Bang. The universe has always existed, and will exist for ever. As old galaxies disappear over the boundary of the observable universe, new ones are created from matter which appears in the form of hydrogen atoms. These atoms come into being spontaneously, from nothingness. The rate of creation is far too slow to be measured (any more than one could easily track down a new grain of sand on Bognor Beach), but space is so large that from this point of view the theory is not impossible. If it is valid, then the universe has always looked much the same as it does now. If our capsule whirls us back, say, 1,000,000 million years into the past, or wafts us 1,000,000 million years into the future, we will see the same numbers of galaxies, but not the same ones as we do today.

This theory, known as the Steady-State theory for obvious reasons, failed to stand up to criticism. In particular, it would mean that conditions are the same all over the universe. When we look at a very remote object such as a quasar, we are in effect looking backwards in time, and we are seeing the universe as it used to be when it was young. On the Steady-State theory, the distribution of galaxies and quasars would be the same as it is nearer home. Radio astronomers soon proved that this is not the case, and, rather sadly, the Steady-State theory was cast aside.

There remains the Oscillating Universe theory, according to which there are Big Bangs every 80,000 million years or so. After each Bang, the universe expands, but the expansion does not continue indefinitely. The galaxies and quasars slow down, stop, and then rush together again, so that there is a new explosion and the entire cycle is repeated. If this is true, then we can still assume that the universe is infinitely old.

This time we can check, because everything depends upon whether the present expansion of the universe can continue until all the groups of galaxies have lost touch with all the other groups. The key is given by the average density of material in the

NGC 2997. A typical spiral galaxy – one of many which we have seen during our journey; we may say that the galaxies are the essential building-blocks of the universe we know.

universe. If it is greater than one atom of hydrogen per cubic metre, then there is enough mass to pull the galaxies back from the brink. If the density is less than this critical value, the galaxies will not return; they will 'escape' from each other, and there will be no new Big Bang. It has been commented that on the Oscillating theory, the universe is like a clock that is periodically rewound; on the alternative theory it is like a clock that can never be restarted once it has stopped. The universe began at a set moment, is evolving, and will eventually die.

Present evidence seems to indicate that the second picture is correct – but remember, there is still the problem of the missing mass. We found it during our journey to Sagittarius A West in the centre of the Galaxy; we also need extra mass to keep some of the clusters of galaxies in a stable condition. The missing mass may be found in unexpectedly high densities of some of the interstellar clouds. It may be due to neutrinos, which are curious particles with no electrical charge and virtually no mass (the Sun emits plenty of them). It has been thought that the mass of a neutrino is nil: but if this is not the case, then the added mass due to neutrinos might push the density above the critical level. And, of course, we can always assume that the missing mass is locked up in Black Holes.

So let us go back into our capsule, and prepare for our last journey, this time into the future. If we hover near the Solar System, there will be no major change for at least 5,000 million years, and probably longer, but when the change comes it will be dramatic. We see that something is happening to the Sun. It has exhausted its supply of available hydrogen fuel; its interior shrinks and heats up, while its outer layers expand, and the smallish, yellow Sun changes into a huge, bloated, red Sun with a cooler surface but a much greater total output. Mercury and Venus suffer first, and are destroyed. The Earth is at the limit of the main danger zone, but we see the oceans start to boil; soon the water evaporates, the atmosphere is stripped away and the Earth is left as a barren, scorched globe from which every trace of life has been wiped out. Mars may survive, but it, too, will lose what remains of its atmosphere. The giant planets and their satellites will be warmed, but there is no chance that they will become habitable, and

Impression of the interstellar ship, capable of travelling to the stars at nearly the speed of light by scooping up hydrogen atoms from space with a huge collector. These would be accelerated into the fusion engine and ejected from the exhaust.

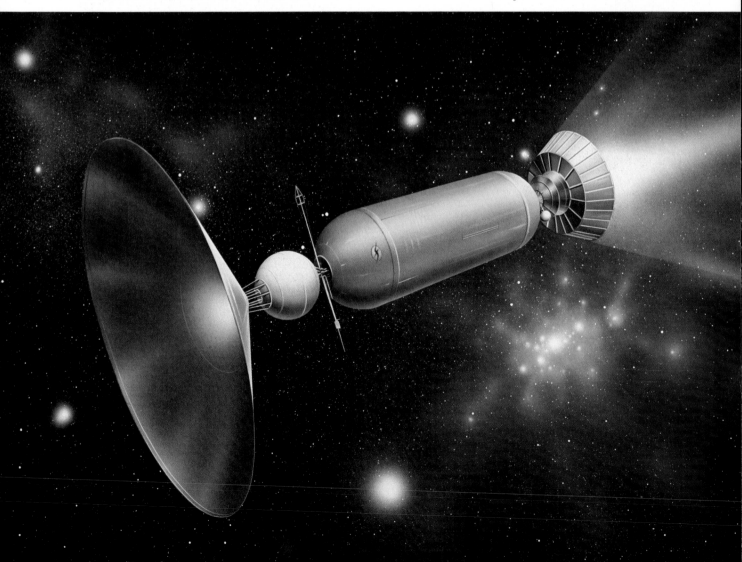

NGC 6164–5, though similar to a planetary nebula, is in fact rather an unusual object. The central star known as HD 148937, is the brightest member of a triple system of stars in orbit around each other.

in any case the Sun's red giant stage will not be prolonged. We see the outer layers thrown off, and the Solar System becomes a planetary nebula, with a central, dying Sun, a shell of gas and a ghostly retinue of the remaining planets. In only a few millions of years the shell of gas has expanded so greatly that it disappears; it has merged with the interstellar medium, while the Sun contracts into the white dwarf condition and the giant planets are again frozen. The Solar System is bitterly cold; it has come almost to the end of its story. White dwarfs shine feebly for a long time, perhaps for 10,000 million to 15,000 million years, but they

cannot last for ever, and the final fate of the Solar System is sealed. The Sun loses the last of its light and heat, and becomes a dead globe – a black dwarf.

With other stars, things are more hurried. From our capsule we see the luminous giants explode as supernovæ or collapse into Black Holes; Rigel, with its 60,000 Sun power, cannot remain on the Main Sequence for more than about 400,000,000 years, and the real cosmic searchlights, such as Eta Carinæ and S Doradûs, are even shorter lived. Again and again the explosions occur, and again and again fresh stars are produced from the ejected material, but the Galaxy itself is dying; the supply of energy is not inexhaustible.

Looking around from our capsule several thousands of millions of years hence, we see

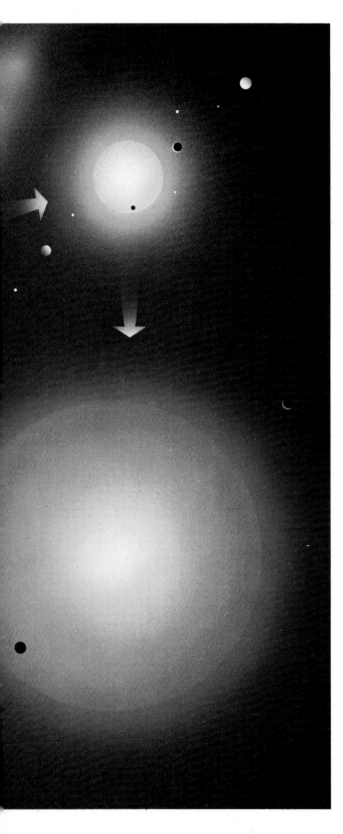

End of the Solar System. We can follow the evolution of the Sun, from its formation out of a huge cloud of gas and dust to its red giant stage before finally collapsing and then shrinking to a white dwarf about 5,000 million years from now.

that the sky has changed. There are still stars, and we can still make out the Andromeda Spiral, the Magellanic Clouds and other members of the Local Group, but where are the Virgo and Coma clusters and their kind? The expansion has continued; they are much further away than they were at the start of our journey. If we watch for long enough, we will come to a period when every group of galaxies has fled over the observable limit. The Local Group is cut off; there can be no way of getting in touch with any systems beyond it. And eventually all the stars in our Galaxy, M.31, Maffei 1 and the other members of the group will cease to shine.

We can take matters still further. It is known that energy will always pass from a hot body to a cooler one. Thus, if you fill two containers with water, one hot and one cold, and then allow them to mix, the result will be water that is at an even temperature. It is a fundamental law that the total energy of the universe must always remain the same; over sufficient time, then, the temperatures will even out. During the last century, the German physicist Rudolf Clausius introduced the term 'entropy'; the lower the entropy, the more uneven is the energy distribution, while higher entropy means a more even distribution. The entropy content of the universe is increasing all the time. Eventually it will reach its maximum, and all change will stop. The universe will still exist, but it will be static and lifeless. We may even compare it with a cine film that is suddenly halted, so that we are left with only one unchanging frame.

It is a gloomy picture, but it is highly speculative, and we may have overlooked some all-important fundamental point which will change the picture completely.

Moreover, the 'heat death' applies only to the Big Bang theory. If the universe is in an oscillating state, or if the Cambridge team can be proved to have hit on the truth after all, there may well be an infinite future. If we could linger in our capsule for long enough, we would find the answer; as things are, we can only guess. But at least one fact is inescapable. If we watch the Solar System for another few thousands of millions of years we will witness the death of the Earth and most of the planets, followed by the end of the Sun. Even if the universe is eternal, the Solar System is not.

Our voyages in imagination are over. We have crossed the universe at the velocity of light, taking with us our Super Telescope, which can show the Earth in detail even across vast reaches of space; we have travelled in a time capsule from the Big Bang of 15,000 million years ago up to the remote future, when the Sun and the Earth will have died, and the universe is ancient. Yet we must bear in mind that though this sort of journey is intriguing, it can never be actually made. Finally, let us come back to reality, and see what the future may hold.

In 1840 a famous English scientist, Dr Dionysius Lardner, was proclaiming that crossing the Atlantic by steam-power was just as impossible as reaching the Moon. A French philosopher, August Comte, instanced the chemistry of the stars as one problem which mankind could never solve. The sceptics are still in full cry. One of America's leading astronomers of 80 years ago, Simon Newcomb, proved to his own satisfaction that no heavier-than-air machine could ever fly, disregarding the fact that Orville and Wilbur Wright had already made successful tests of their aeroplane. In 1934 Sir Archibald Sinclair, then British Under-Secretary of State for Air, wrote that with regard to jet propulsion, 'scientific investigations into the possibilities has given no indication that this method can be a serious competitor to the airscrew-engine combination; we do not consider that we would be justified in spending any time or money on it.' A year later Forest Ray Moulton, yet another leading American astronomer, dismissed all talk of flights to the Moon: 'There is no theory which would guide us through interplanetary space to another world even if we could control our departure from the Earth . . . and there is no way of easing our ether ship down on to the surface of another world, even if we could get there.' This was also the opinion of many popular writers as recently as 1948. On 18 October of that year a columnist on the London *Daily Mirror* commented that 'our candid view is that all talk of going to the Moon, and all talk of signals from the Moon, is sheer balderdash – in fact, just moonshine'. Little more than twenty years later, Neil Armstrong stepped out on to the Sea of Tranquillity.

We know that travel within the Solar System is possible. The Moon and Mars at least can be colonized to some extent, and automatic probes can contact all the other planets, even out to Pluto and Neptune; it is quite on the cards that astronauts will follow. Yet rockets have their limitations. Those of today are so slow that they would take immensely long periods to reach other stars; thus, Voyager 1, now on its way out of the Solar System, is not expected to be close to another star until about AD 40,000, when it ought to be within range of a faint star in the constellation of the Little Bear. There is no conceivable way in which we could maintain contact with a probe of this sort, and the idea of sending a human crew is frankly absurd.

All kinds of dodges have been suggested. Perhaps the astronauts of an interstellar ship could be put into a state of suspended animation, so that they would remain unconscious and unchanging for most of the journey and would be woken up just in time

The world's largest dish-type radio telescope is at Arecibo in Puerto Rico. It has been built in a natural basin and therefore cannot be steered, but it is extremely powerful.

to make a convenient landing upon a planet of Barnard's Star or Delta Pavonis? Could there be an interstellar ark, totally self-supporting, so that the original crew members would die early in the trip and only their descendants would survive to reach their chosen destination? Could we produce a race of men (and women) with immensely long life-spans, so that they could cope with journeys lasting for thousands of years? Candidly, I feel that all ideas of this kind belong strictly to the realm of science fiction. With our present technology, reaching planets of other stars is impossible, and would be impossible even if we could be

sure of the exact positions of inhabited planets in other Solar Systems.

Note that I say 'with our present technology'. Without this qualification I would be falling into the same trap as Dr Lardner, Professor Moulton and the anonymous columnist of the *Daily Mirror*. We need a fundamental breakthrough, simply because if we are to achieve interstellar flight we must overcome the limitations imposed by the velocity of light. When we discuss thought-travel, teleportation and the like, to say nothing of space-warps and time-warps, we are still in the realm of science fiction – but are such exotic methods any more unlikely than television, or even radio, would have seemed to Julius Cæsar? The breakthrough may come this year, next year, in a century's time, in a thousand or a million years – or never. Unless or until it does, we must confine ourselves to travel within the Solar System.

Of course, this does not rule out the slim possibility of picking up artificial signals; remember Tau Ceti and Epsilon Eridani. The world's largest radio telescope, the non-steerable 1,000-foot dish built in a natural hollow at Arecibo in Puerto Rico, has already been used to beam a message to the globular cluster in Hercules, though since it will not arrive there until AD 25,000 we can hardly hope for an early reply. Even if we did establish that other intelligent races exist, we could do little more than make our presence known to them, and even at 186,000 miles per second we must admit that radio waves are inconveniently slow.

We must also consider the future conditions on the Earth itself. Over the past few thousand years, Man has progressed from the cave-dwelling stage to the Space Age; the developments over the past few decades have been particularly striking. If we continue to advance at this rate, there is no knowing what we may achieve in the foreseeable future, and by AD 10,000 the science of 1983 will appear as antiquated as the invention of the wheel does to us. Unfortunately, we also have to reckon with politics, which represent the very worst side of human nature. Before 1945, wars were largely confined to professional armies, and the results did not matter much in the long run; thus, if Carthage had destroyed Rome instead of vice versa, you and I would still be here. Now, of course, things are different, and a nuclear war would not only destroy civilization but would quite possibly make the Earth permanently uninhabitable. It may well be that every civilization, wherever it appears, goes through this testing period, when it has the scientific knowledge to destroy itself and lacks the culture to refrain from doing

so. Pessimists even suggest that this is the norm, in which case there must be many ruined, radioactive planets not only in our Galaxy but also in others. We can only hope that we do not add one more to a list of failures. If we avoid nuclear conflict for another two or three centuries we will probably have learned enough to avoid it permanently. The choice is ours.

Lastly, what are the prospects of the Earth being visited by advanced beings from a distant system? Stories about flying saucers, little green men from Mars and ancient astronauts are great fun, but they are no more than that, and are strictly for consumption by the credulous only. This is not to say that such contacts are impossible. A visit from a flying saucer cannot be ruled out, even though there is not the slightest evidence that it has happened yet. Remember, too, that a visitor who came here a few million years ago would have found a world at a very primitive stage, and would have been unlikely to linger; he would have reported that there was no advanced life on Planet Three of the Sun. Yet a few million years is a very short period by cosmical standards.

Some astronomers have warned that sending out radio messages to advertise our presence is dangerous, because of the risk of invasion and takeover by some super-civilization. To me, this seems sheer nonsense. No civilization could reach us unless it had survived its trial period and had achieved real enlightenment, in which case we need not fear it – indeed, it could certainly teach us a great deal. No warlike race could ever conquer interstellar flight, and this applies equally to ourselves.

Perhaps this is the right moment to stop. I hope you have enjoyed our travels. We are part of a great universe, and if life really is rare, as some people believe, then we may not be quite so unimportant as we often think.

Glossary

Albedo The reflecting power of a planet or other non-luminous body.

Aphelion The point of furthest recession of a planet (or other body) from the Sun.

Apparent magnitude The apparent brightness of a star (or any other celestial object). The lower the magnitude, the brighter the object. The faintest stars normally visible with the naked eye on a clear night are of magnitude 6.

Asteroids The minor planets, moving mainly in the zone between the orbits of Mars and Jupiter, though some depart from the main zone.

Astronomical unit The distance between the Earth and the Sun (approximately 93,000,000 miles or 149,600,000 kilometres).

Becklin-Neugebauer Object (BN) An infra-red source inside the Orion Nebula. It is either a very powerful star, or else a star in the process of formation.

Binary star A stellar system made up of two stars which are genuinely associated.

Black Hole A region round a very small, very massive, collapsed star from which not even light can escape.

Caldera An enlarged volcanic crater.

Cepheid A short-period variable star, useful because there is a definite relationship between the period of variation and the real luminosity, so that the distance of the star may be found.

Chromosphere That part of the Sun's atmosphere immediately above the bright surface or photosphere.

Corona The outermost part of the Sun's atmosphere.

Cosmic rays High-velocity atomic particles reaching the Earth from outer space.

Doppler Effect The apparent change in wavelength of the light of a luminous body in motion relative to the observer. With an approaching object the wavelength is shortened (blue shift); with a receding object, lengthened (red shift).

Electron Part of an atom; a fundamental particle carrying a negative electric charge.

Escape velocity The minimum velocity that an object must have in order to escape from the surface of a planet, or any other object, without being given any extra impetus.

Event horizon The boundary round a Black Hole inside which the gravitational effects of the old, collapsed star prevent even light from escaping.

Faculæ Bright, temporary patches on the surface of the Sun.

First magnitude 'Magnitudes' are grades of the apparent brightness of stars. The system works rather in the manner of a golfer's handicap, with the more brilliant performers having the lower values. Thus, stars of magnitude 1 are bright. Conventionally, stars of from magnitude −1.4 (Sirius) to +1.3 (Beta Crucis) are classed as being of the first magnitude.

Flares, solar Brilliant eruptions in the outer part of the Sun's atmosphere.

Flare star Faint red dwarf stars which show sudden, short-lived increases in luminosity.

Galaxies Systems made up of stars, nebulæ and interstellar matter.

Galaxy, the The Galaxy of which our Sun is a member.

Globules Small, dark patches inside gaseous nebulæ, possibly embryo stars.

H.I and H.II regions Clouds of hydrogen in the Galaxy. In H.I regions the hydrogen is neutral; in H.II regions the hydrogen is ionized. The presence of hot stars will make the cloud shine as a nebula.

Hertzsprung-Russell (or H-R) Diagram A diagram in which stars are plotted according to their spectral types and their real luminosities.

Infra-red radiation Radiation with wavelength longer than that of visible light.

Interferometer, stellar An instrument for measuring star diameters. The principle is based upon light interference.

Ionosphere The region of the Earth's atmosphere lying above the stratosphere.

Light-year The distance travelled by light in one year: approximately 5,880,000 million miles.

Local Group A group of more than two dozen galaxies, of which our Galaxy and the Andromeda Spiral are members.

Main Sequence star A star in a stable condition, radiating because of the conversion of hydrogen into helium. The Sun is a typical Main Sequence star.

Mass The quantity of matter that a body contains.

Meteor A small particle moving round the Sun, visible only when it enters the upper air and is burned away.

Meteorite A larger object which can reach the ground without being destroyed. Meteorites are not large meteors; they are more nearly related to the asteroids.

Nanometer A unit for measurement of very short lengths e.g. the wavelength of light. It is equivalent to one thousand millionth of a metre or 10 Ångströms.

Nebula A cloud of gas and dust in space.

Neutrino A fundamental particle with no electric charge. It was formerly believed to have no mass, though recent investigations indicate that it may have a little.

Neutron A fundamental particle with no electric charge, but a mass almost equal to that of a proton.

Neutron star The remnant of a very massive star that has exploded as a supernova. Also known as a pulsar.

Nova A star that suddenly flares up temporarily to many times its normal brightness. All normal novæ are binary systems.

Orbit The path of a celestial object.

Oscillating Universe theory The theory that the universe will eventually contract to produce a new 'Big Bang'. On the Oscillating Universe theory, Big Bangs will occur at intervals of about 80,000 million years.

Parallax The apparent shift in position of a relatively near-by star against the background of more distant stars, due to the Earth's motion round the Sun over a period of six months.

Perihelion The point of closest approach of a planet (or other body) to the Sun.

Photon The smallest 'unit' of light.

Photosphere The bright surface of the Sun.

Planetary nebula A small, dense, hot star surrounded by a shell of expanding gas.

Prominences Masses of glowing gas rising from the surface of the Sun, made up chiefly of hydrogen. They are visible with the naked eye only during a total eclipse.

Proper motion, stellar The individual movement of a star.

Proton A fundamental particle with a positive electric charge.

Pulsar A neutron star, spinning rapidly and emitting radio waves.

Quasar A very remote, super-luminous object – possibly the central part of a very active galaxy.

Red dwarf A dim red star.

Red giant A very large, luminous red star.

Schwarzschild radius The radius that a body must have if its escape velocity is to be equal to the velocity of light.

Seyfert galaxy A galaxy with a small, bright nucleus and weak spiral arms.

Solar System The system of which the Sun is the most important member.

Spectrum The splitting up of light into its constituent colours, by a prism or some equivalent device.

Steady-State theory The theory that the universe has always existed and will exist for ever. It is now generally rejected.

Supernova A very massive star which suffers a cataclysmic outburst, ending its career as a patch of expanding gas with (often) a neutron star remnant.

Synchronous rotation If the rotation period of a satellite of a planet is the same as the sidereal period (i.e. the time taken to complete one orbit), the rotation is said to be synchronous (or 'captured'). Most of the planetary satellites, including our Moon, have synchronous rotations.

Synchrotron emission The emission of radiation by charged particles moving in a magnetic field.

Transfer orbit The path followed by a body moving between the orbits of two planets, without being given extra impetus.

Van Allen Zones Zones of charged radiation round the Earth.

Variable stars Stars that alter in brightness over relatively short periods.

White dwarf A very small, very dense star that has used up its nuclear energy, and is in a very late stage of its evolution.

ZAMS Zero age main sequence.

Zenith The observer's overhead point.

Zodiac A belt stretching round the sky in which the Sun, Moon and bright planets are always to be found.

Zodiacal light A cone of light rising from the horizon after sunset or before sunrise. It is due to thinly spread material in the main plane of the Solar System.

Illustration Acknowledgments

The author and publishers are grateful to the individuals and organizations listed below for permission to reproduce copyright material:

The Anglo-Australian Telescope Board, photos by D. F. Malin 127 © 1980, 134 © 1981, 158–9 © 1981, 175 © 1979, 179 © 1980, 181 © 1981; The Association of Universities for Research in Astronomy Inc. 110, 148–9; The Association of Universities for Research in Astronomy Inc., The Cerro Tololo Inter-American Observatory 83, 165; The Association of Universities for Research in Astronomy, Inc., The Kitt Peak National Observatory 142, 177 photo by Gabriel Martin; Jack Bennett 74; BBC photo by Pieter Morpurgo 84; California Institute of Technology and Carnegie Institution Washington 8, 62; The Hale Observatories, Mount Wilson 41; Commander Hatfield 81; Jodrell Bank, Cheshire 87; Kennedy Space Center 95, 118; The Kitt Peak National Observatory 172; Mauna Kea Observatory 94; Patrick Moore 37, 80, 98, 109, 150, 151, 170–71, 185; NASA frontispiece, 10, 11, 12, 13, 16, 19, 20, 23, 24, 25, 26, 27, 32, 33, 34–5, 40, 42, 44, 45, 46, 50, 51, 52, 53, 54, 55, 56, 57, 58, 59, 60, 61, 63, 64, 65, 66, 68, 75, 82, 86, 92–3, 104, 114, 115, 119, 135, 146, 156, 162, 168; Novosti Press Agency, London 90; Palomar Observatory 113, 130, 147, 157, 176; Royal Observatory Edinburgh half title © 1982 photo by David Malin AAT, 88 © 1979 photo by David Malin AAT, 101 photo by David Malin AAT and Photolabs, 108 © 1979 photo by Photolabs, 121 © 1978, 152 © 1978; the Smithsonian Institution 164 © 1981; Dr Hans Vehrenberg 126.

The author and publishers would particularly like to thank Paul Doherty for the artwork and diagrams reproduced on pages 6–7, 14, 15, 28–9, 31, 36, 38–9, 48, 70–71, 73, 76–7, 79, 91, 96–7, 99, 103, 107, 116, 120, 122–3, 124–5, 128–9, 133, 137, 138–9, 140, 144, 155, 161, 166, 169, 180, 182–3; and are grateful to Frances Rowsell for her assistance in picture research in the United States.

If in any case the acknowledgment proves to be inadequate the publishers apologize. In no case is such inadequacy intentional, and if any owner of copyright who has remained untraced will communicate with the publishers, the required acknowledgment will be made in future editions of the book.

Index